THE TECHNICAL HISTORY OF CHINA'S GRAND CANAL

THE TECHNICAL HISTORY OF CHINA'S GRAND CANAL

Xuming Tan
Yunpeng Li
Jun Deng
Jinhong Wan
Jiangang Liu

China Institute of Water Resources and
Hydropower Research, China

中国水利水电出版社
China Water & Power Press

World Century

Published by

World Century Publishing Corporation
27 Warren Street, Suite 401-402, Hackensack, NJ 07601, USA

Library of Congress Cataloging-in-Publication Data
Names: Tan, Xuming, author.
Title: The technical history of China's Grand Canal / Tan Xumin.
Description: [New Jersey] : World Century, [2019]
Identifiers: LCCN 2019020443 | ISBN 9781945552038
Subjects: LCSH: Hydraulic engineering--China--Grand Canal. |
Natural history--China--Grand Canal. | Grand Canal (China)--History.
Classification: LCC TC702.G7 T36 2019 | DDC 627.0951/156
LC record available at http://lccn.loc.gov/2019020443

British Library Cataloguing-in-Publication Data
A catalogue record for this book is available from the British Library.

For any available supplementary material, please visit
https://www.worldscientific.com/worldscibooks/10.1142/U033#t=suppl

Desk Editor: Ser Kei Vong/Ong Shi Min Nicole

Typeset by Stallion Press
Email: enquiries@stallionpress.com

Contents

About the Authors

Tan Xuming is Vice-Chief Engineer of China Institute of Water Resources and Hydropower Research; Member of Working group on History of Irrigation, Drainage and Flood Control (WG-HIST) of International Commission on Irrigation and Drainage (ICID); Chairman of Committee on Water History of Chinese Hydraulic Engineering Society. Her research interests include history of water conservancy, history of canal technology, flood control, irrigation, river basin water environment evolution, and disaster and disaster reduction.

Li Yunpeng is Director of Water History Department of Research Center on Flood and Drought Disaster Reduction of the Ministry of Water Resources; Member of Working Group on History of Irrigation, Drainage and Flood Control (WG-HIST), International Commission on Irrigation and Drainage (ICID). His research interests include history of water structure and water environment, and water heritage protection.

Deng Jun is Senior Engineer of China Institute of Water Resources and Hydropower Research. She mainly researches the water conservancy history and water culture, programming and management of water resources, and water resource accounting.

Wan Jinhong is Associate Professor of China Institute of Water Resources and Hydropower Research. He is mainly engaged in the study of the history of disasters, hydro-social history, and water culture.

Liu Jiangang is Senior Engineer of China Institute of Water Resources and Hydropower Research (IWHR). His research interests include water conservancy history and water culture, water heritage protection, water resources management and planning, and flood and drought disaster law analysis.

Preface

The Grand Canal and the Great Wall in China are the landmark projects of the Chinese civilization. From the Warring States to the early 20th century, in the historical period up to 2,500 years, the Grand Canal has been continuing construction and applications for water transportation across the major rivers to the final destination of the capital. The project remains a deep imprint of the political, economic, technological, cultural patterns during various historical periods and demonstrates the unyielding national spirit and great creativity of the Chinese.

In the process of Chinese civilization, countless canals were dug during the various historical periods. However, independent engineering projects called the Grand Canal were built up through continuous construction and usage using hundreds of years or even thousands of years. There are two such projects. The first Grand Canal consisted of the Tongji Canal (or Bian Canal), the Yongji Canal (or Weihe River), Huaiyang Canal, Jiangnan (the south of the Yangtze River) Canal, and Zhedong (the eastern Zhejiang) Canal during the Sui, Tang, and Song Dynasties, with Luoyang or Kaifeng as the centre. The construction of this Grand Canal started with the canals between the Yangtze River and Huaihe River as well as between the Yellow River and Huaihe River (i.e. Hangou and Honggou canals) and lasted for nearly 1,000 years. The other Grand Canal starting from the Yuan, Ming, and Qing Dynasties, consisted of Tonghui Canal, the North Canal, the South Canal, Huitong Canal, Huaiyang Canal, Jiangnan Canal, and Zhedong Canal, from Beijing to Hangzhou and to Ningbo, with a total length of 2,000 km. The canals were continuously used for 600 years. All these canals together are called the Chinese Grand Canal.

From the north to the south, the Chinese Grand Canal connects the Haihe River, the Yellow River, the Huaihe River, the Yangtze

River, the Qiantang River, and the Yongjiang River, covering most of eastern China. Average annual rainfalls in the different regions along the Grand Canal vary from 500 to 1,500 mm. The topography and water resources are dramatically different in these regions. The complexity of engineering problems involved, and the tremendous inputs of manpower and materials to build the Grand Canal were incomparable with those of other canals in the world. Water management practices with the Grand Canal are also globally unique. Therefore, during different historical periods, many water conservancy projects were built along the canal, with rich construction types and distinctive regional characteristics. These outstanding water works and water transportation projects represent the highest achievement of ancient Chinese water conservancy planning and constitute the core value of cultural heritage of the Chinese Grand Canal.

The Tonghui Canal locates in the semi-arid region of the north China plain. Relying on the water source project (with springs) in Beijing Xishan, Tonghui River functioned as the transportation waterway for more than 600 years, and generated a river and lake network to create a beautiful water environment around Beijing. With the largest amount of sediment in the world, The Yellow River has the characteristics of easily silting up and frequently changing water courses, greatly interfering the canal sections dependent on its water source. During the early Southern Song Dynasty, the Yellow River crossed Huaihe River and began to move to the south for 700 years. The huge amount of silt from the Yellow River buried the Tongji Canal. The Yellow River also changed the natural environment along the banks, which greatly affected the canal sections in Shandong and Huai'an. In order to solve the interference problem of the Yellow River sediment on the canal and to guarantee water transport of goods to the capital, the water conservancy project for management and construction of the Yellow River and Huaihe River continued for as long as 600 years, which produced the Hongze Lake, the fifth largest freshwater lake in China. In the southeast coastal Zhejiang–Ningshao plain, the Sanjiang Gate was constructed in the 16th century, which was composed of 28 sluices with the function to stop salty water and to store fresh water. Therefore, the Sanjiang Gate became the sole water control project of the Grand Canal. Xiaoshan and Shaoxing sections of the Zhedong Canal therefore became an independent engineering system.

The interval canals between two natural rivers evolved to the Grand Canal to connect multiple water basins. The Grand Canal formed a continuous water channel, played a decisive role in water transportation for the national economy, and reflected the state wills in different historical periods to control and distribute the water resources. To guarantee goods transportation to the capital through the canal, water resources along the canal were collected to the so-called 'water tanks' through engineering measures, which was the earliest water transfer in the world. The Grand Canal witnessed the paramilitary shipping management practices under the centralized power in different periods. The construction and operation of the Grand Canal for more than 2,000 years resulted in the unique canal engineering system in China with complex hydrological conditions in different regions. The canal system included the oldest reservoirs and complete dam construction in the world, crossed watersheds and water transport facilities of rivers. The water source projects, dams, sluices, and warehouses along the Grand Canal effectively supported the longest continuous waterway in the world. The Grand Canal reflected the tremendous achievements of outstanding Chinese water conservancy science and water management technology before the 17th century.

In early 20th century, with the termination of the Grand Canal's inherent mission (to transport goods to the capital), the canal and its water conservancy facilities changed greatly. Currently, the section of the Beijing–Hangzhou Canal from Jining to the south still retains the transportation function with a navigable length about 800 km. The other 1,200 km of the canal are still used for river flood control, irrigation, and ecological and landscape functions. In the planning of the east route of South-to-North Water Transfer Project, 90 per cent of the section of Beijing–Hangzhou Canal from Huaihe River to the north will be utilized. In the past 100 years, due to the lack of systematic management, the water environment along the canal was degraded and engineering facilities have been seriously damaged. In order to effectively protect the Grand Canal's cultural heritage, in 2006 the State Council announced the whole section from Beijing to Hangzhou of the Grand Canal as the national key cultural relics protection unit. The State Administration of Cultural Heritage will include the Grand Canal in the 2014 tentative list for declaration of the world cultural heritage. The project to protect the canal cultural heritage has started. Since 2006, the State Bureau of Cultural Relics has supervised the archaeological excavation of some key areas of the Grand Canal and the protection planning of the Grand Canal cultural heritage. As a member unit to prepare the nomination process to apply Grand Canal as the world cultural heritage, the Ministry of Water Resources has made arrangement for protection planning of the ancient water conservancy engineering, protection and management planning (water conservancy project) of the section from Beijing to Hangzhou of the Grand Canal.

The study of the canal heritage in this book was carried out according to the historical background during different periods of the construction and operation of the Grand Canal. Systematically from the river's natural history, engineering and technical perspectives, we studied the canal management practices in different periods, as well as the interrelations between the Grand Canal and the Chinese politics, economics, and culture. Furthermore, we assessed the effects of the Grand Canal in the progress of Chinese civilization, engineering technology achievement, the natural environment, and the society. This book was compiled based on the relationships between the canal and natural rivers. Undoubtedly the core values of the Grand Canal are the historical and cultural values, while engineering and technology provide the key support to demonstrate the cultural value. These show the content of the Grand Canal heritage and the scientific value of the key canal projects. In the last part of this book, we assessed the Grand Canal's core values based on the evaluation criteria for the world cultural heritage, the natural characteristics of each canal section, and on summary of the heritage content. This book fills the following gaps: the definition of Grand Canal, evolution of Tongji Canal and Yongji Canal, interrelationships between the canal and the

Yellow River, Huaihe River, and Haihe River, the integrity of the canal cultural heritage, and the authenticity of the basic factors.

The available data of the Grand Canal today was from the planning work beginning in the 1920s. Especially in the 1930s, Mr. Wang Huzhen was in charge of the Canal Management Committee. Early planning work (survey and measurement) of the Grand Canal was conducted during the period with domestic troubles and foreign invasion. The Grand Canal topography is so far the only one available. This book was based on the accumulation of the Institute of Water Conservancy History during the past 30 years. The paper *Nanwang Water Control Project of Bejing-Hangzhou Canal during the Ming and Qing Dynasties* by Yao Hanyuan (1985) was groundbreaking to study the history of the Grand Canal engineering and technology. Based on solid data and rigorous proofs, this paper elaborated the forming process of the Huitong River, Daicun Dam, and Nanwang water control project, and analysed key engineering and operation mechanism of the Beijing–Hangzhou Canal. Another influential paper during this period was *The History of Beijing Canal Transport and City Water Supply and Drainage* by Cai Fan. The author systematically studied the Tonghui River engineering system, and revealed the technical achievement of water engineering and its profound influence on the development of Beijing city from the regional natural environment and the characteristics of water resources, scientific planning, and engineering and technical perspective. Other basic works in the 1980s included the annotation of *(Ming) Caohe River Atlas* and *Twenty-five Histories of Rivers and Canals.* They are the important foundation of historical data. Especially *(Ming) Caohe River Atlas* was the first monograph about the Grand Canal, most of which has been lost. The Chinese Water Conservancy History Research of Japan provided us with a copy of the book. Based on the copy and three incomplete volumes of the proofread *Siku Quanshu*, we finished this book in Chinese. During the 1990s to the early 21st century, research monographs about the canal engineering and technology included *History of the Beijing-Hangzhou Grand Canal*, *History of Chinese Water Conservancy*, *History of Water Conservancy and Transportation*,

Chinese Science and Technology (Volume of Water Conservancy), and *Hongze Lake Reservoir Construction*. The previous work provides important foundation for today's research and protection of the Grand Canal cultural heritage.

Since 2007, cooperating with researchers from the Chinese Academy of Cultural Heritage and Tsinghua University, we conducted several interdisciplinary scientific investigations of the Grand Canal. From these investigations, we obtained the first hand information to understand the Grand Canal engineering and management, as well as the current situation of the cultural heritage, which should be greatly important to carry out research and protection for the cultural heritage.

Here, we pay tribute to the predecessors represented by Mr. Wang Huzhen and Mr. Yao Hanyuan, who had carried out strenuous work to save the Grand Canal. We are grateful to the colleagues working with us for the study of the Grand Canal heritage.

Chapter 1
Introduction

1.1 Introduction

The construction of the Chinese Grand Canal began in the late Spring and Autumn Period. It was the product of territorial expansion by the vassal states. Due to political and military demands, artificial waterways were built and continuously improved, forming the genesis of the Grand Canal itself. Since the unification of China's political system after the Qin and Han dynasties, the canal has represented a transportation lifeline that maintains political and economic connections to central government and a large-scale water-conservancy project under national management. Political, economic, cultural, and technological developments over various historical periods have given the cultural value of the canal a specific historical context. The unique characteristics of the natural environments surrounding the canal and its river hydrology have provided opportunities for the creation of engineering technology, resulting in distinctive regional projects and management patterns.

Based on the available research results, we can define the Chinese Grand Canal as a continuous waterway that stretches for more than 1,000 km, with a relatively independent engineering system. It is also an artificial waterway system consisting of many river reaches. According to this definition, the Chinese Grand Canal generally refers to the Beijing–Hangzhou Grand Canal and the Sui–Tang Grand Canal.

A canal is an artificial waterway that connects two natural rivers. Canals were continuously constructed and improved to form multiple connecting rivers; these eventually became an independent engineering system constituting a waterway that saw continuous traffic. The canalised natural watercourses connected to the canal, such as the Zhedong Canal, were regarded as part of the extension of the canal. The water engineering, hydraulic structures, water facilities, and transport management across the different reaches of the Grand Canal show significant differences because of the differing natural environments of the regions along the canal.

In order to analyse the cultural value of the Grand Canal, it is essential to study its construction, development, and technical characteristics in terms of natural and water resources. The termini of the canal have changed throughout the canal's history, and therefore the canal's engineering and management has been influenced by these different environments and the political, economic, scientific, technological, and cultural aspects of the areas through which it travelled.

In order to understand the significant cultural heritage values of the canal, it is important to begin by understanding the canal's morphology and engineering technology across these different periods and locations. The rest of this chapter will be dedicated to the natural background and history of the Grand Canal.

1.2 The Natural Conditions of the Grand Canal

1.2.1 *The Impact of Topography on the Distribution of the Grand Canal*

The overall altitude of Chinese terrain changes from the northwest to the southeast. Mountainous areas, plateau basins, hills, and plains account for approximately 33 per cent, 26 per cent, 19 per cent, and 12 per cent of the total area of the country, respectively. China's mountains mostly stretch from east to west and northeast to southwest. These mountain ranges determine the direction in which the rivers flow. Since most of China's major rivers flow into the sea from the west to the east, the water system separated by the geographical environment is the natural background of the Grand Canal from the north to the south.

The Grand Canal approaches and crosses the North China Plain, the Yangtze River Delta Plain, and the eastern part of the Zhejiang Xiaoshao Plain from different directions. The region south of the Yangtze River has a warm and humid climate, fertile land, and abundant resources; it is China's major agricultural economic zone.

During the dynasties, the main destinations of grain transportation were the capitals, and waterways were utilised to achieve the connection between these economic and political centres. From the Sui and Tang dynasties to the Qing Dynasty, the final destinations for grain along the Grand Canal were Chang'an, then Luoyang, then Kaifeng, and, finally, Beijing. The canals connected the water system of the Haihe River Basin and connected the Yellow River and the Huaihe River. Connections were also made between the Yangtze River, Lake Taihu, and the Qiantang River. The Chinese Grand Canal is the longest in the world, crossing the most natural rivers. The separation of the political and economic centres in each historical period was the social factor that affected the canal's wide pattern of distribution.

In order to connect the rivers with the Grand Canal, the problem of terrain elevation — which resulted from watersheds of drainage basins and different rivers — had to be addressed. Before the advent of fossil fuel in the 17th century, the Beijing–Hangzhou Grand Canal was the canal that had the highest elevation difference between its two termini. The regional conditions of hydrology and water resources were the natural factors that affected water project planning, resulting in various types of water source and the canal's control works.

1.2.2 *The Impact of Hydrology and Water Conditions on the Water-Supply Projects*

China is located in the East Asian monsoon region. The distribution of water resources differs greatly between areas and seasons; from the north to the south, the average annual rainfall over the Yellow River–Huaihe River Plain ranges from 500 to 700 mm, whereas the average annual rainfall from the south of the Huaihe River to the Qiantang River Basin ranges from 1,000 to 1,500 mm. Approximately 60–70 per cent of the annual rainfall in the east is concentrated in the four months from June to September. During the flood season, the rainfall of one month might account for 50–70 per cent of the annual rainfall. The hydrology and water resource conditions cause the serious problem of water resources in China. On the other hand, the spatial and temporal variations of water resource distributions provided the opportunity for the creative engineering of the Grand Canal, from which abundant types of waterworks and unique hydraulic or waterway constructions were produced along the canal.

By reconstructing and interconnecting waterways in the different regions of China, the Grand Canal gradually became a relatively independent engineering system. To obtain water resources, the early canals utilised as many rivers and lakes as possible, meaning the waterways twisted around the landscape. The changes in water level across the rivers and lakes, silt beaches, and tides greatly impacted the operation of the canal. As the importance of water transportation increased, the waterways were gradually straightened, and projects for water diversion and regulation were constructed and gradually improved. The link between a canal section and a natural river finally concentrated on the joint (*yunkou* or 'canal mouth'), and the main body of the canal was separated from the natural river. The joint was the junction of a canal section and a natural river, from which the canal's water source was obtained and the sediment problem was resolved. The joints were also the transportation hubs between rivers and canals. Facilities for water transport included ports and transit warehouses.

Studies about the evolution of the estuaries and coasts have indicated that the low coasts from the Bohai Sea to the East China Sea have changed significantly. Over the past 2,000 years, due to the transportation of huge amounts of sediment, the river mouths of the Haihe River, Yangtze River, and Yellow River have moved forward, reflecting that the canal reaches and joints downstream could have taken more advantage of ocean tides in earlier periods as opposed to today. The joints connected with the Yangtze River and Huaihe River and the engineering and port utilising tides for transportation on the Zhedong Canal were built at least as early as the Eastern Jin Dynasty and became perfected in the Tang and Song dynasties.

The Grand Canal has a significant impact on rivers. The construction of the Yongji Canal cut off the flow paths of the south portion of the Haihe River into the Yellow River, which created the main streams in the southern Haihe River — the Weihe River and the South Canal. Due to the transformation of the northern section of the Yongji Canal, the northern portion of the Haihe River changed its flow path to form the main stream in the north, the North Canal.

1.2.3 *The Impact of the Yellow River on the Canal Projects*

The Yellow River has the largest sediment content in the world. During the late Southern and Northern dynasties (during the 5th century), the influence of the Yellow River caused the lakes along the canals — such as *Ying Ze* (Yingze Lake), *Putian Ze* (Putian Lake), *Daye Ze* (Daye Lake), and *Mengzhu Ze* (Mengzhu Lake), which were used for sediment and transportation — to gradually disappear. The Tongji Canal was built during the Sui and Tang dynasties.

The impact of the flooding of the Yellow River caused the sedimentation problem to become very serious. In order to retain the ability to navigate the canal, tremendous manpower had to be involved in dredging it. During the Tang and Northern Song dynasties, the military was even used specifically for maintenance of the Bianhe Canal. This led to the development of water course structures such as wooden shores, pick dams, and sluices.

For about 700 years, from the 2nd year of Jianyan during the Song Dynasty (AD 1128) to the 5th year of Xianfeng during the Qing Dynasty (AD 1855), the Yellow River flowed to the south by crossing the Huaihe River. Due to sediment accumulation and the frequent changes of the flow paths in the north and south, the flood paths of the Yellow River greatly interfered with the northern and southern canals. The influence of the northern flood paths of the Yellow River caused the gradual separation of the rivers of Sishui, Yishui, and Shushui within the northern part of the Yellow River. These rivers evolved as independent basins, resulting in Weishan Lake, Luoma Lake, and other lakes. The canals to the north of the Yellow River changed paths frequently due to the Yellow River flooding and other natural river changes.

In the 16th century, after 300 years of operation, Xuzhouhong, to the east of Xuzhou, and Lyulianghong, downstream of Sishui, were filled with silt, ending the water transportation along the Yellow River. The northern joint between the Grand Canal and the Yellow River therefore continued moving downstream from Xuzhou to Yangzhuang, in Huai'an.

In the 16th century, the southern flood path of the Yellow River initially buried the canal between the Yellow River and the Huaihe

River (i.e. Bianhe Canal) under sediment. Then, the sediment from the Yellow River clogged the downstream waterway of the Huaihe River. A massive water conservancy project was thus launched along the Qingkou section of the Huaihe River from the late 16th century to the mid-17th century. This water-conservancy project and the Yellow River siltation resulted in several artificial lakes, including the Hongze Lake, in the lower reach of the Huaihe River, and Weishan Lake and Luoma Lake, both in the middle reach of the Huaihe River. The rivers of Sishui, Yishui, and Shushui to the north of the Huaihe River evolved as an independent water system that flowed to the sea.

During the mid-19th century, the downstream flow path of the Yellow River to the Huaihe River and the sea finally came to an end. In 1851, the Huaihe River stopped flowing into the sea and began flowing into the Yangtze River instead, meaning the Huaiyang Canal became the channel from the Huaihe River to the Yangtze River. In 1855, the Yellow River changed its route dramatically and broke the Huitong River between Zhangqiu and Dongying in Shandong.

Social unrest during the late Qing Dynasty and the diversions of the Yellow River meant that water transport along the Grand Canal, which had been taking place for 700 years, came to an end. Due to the division of the water sources — with Jining, Shandong, as the location of the divide — the northern section of the Beijing–Hangzhou Canal was no longer navigable, although the southern section remained passable. The engineering facilities and river features also underwent great changes.

1.3 A Brief History of the Grand Canal

1.3.1 *The Spring and Autumn Period: The Origins of the Canal*

From the late Spring and Autumn Period to the Warring States Period, the merger's wars among the states prompted the construction of state canals to link the natural rivers. Interval canals to connect two rivers or two bodies of water first appeared in the Jianghan Plain (between the Wu and Chu states), which had tremendous strength, flat terrain, and abundant water sources, as well as the Lake Taihu area (between the Wu and Yue states). As canals were mainly constructed for military purposes, the main consideration in the planning of the routes was how to achieve the connection between the rivers with the least amount of excavation and in the shortest possible time (Figure 1.1). These artificial water routes were convoluted and inconvenient when it came to fully utilising the natural rivers and lakes. Once such canals were disregarded, they were quickly filled with sediment. The canals had an important position in terms of transportation and connections between the large rivers, and they were continuously transformed; the artificial waterways were gradually extended, the engineering facilities were improved, and they gradually evolved into completely canalised key waterways across the basins.

Two thousand years ago, Sima Qian (*c.* 145–90 BC) recorded the distribution of the canals during the Warring States Period in *Records of the Grand Historian: Treatises of Rivers and Canals.*

> The Honggou Canal, which connected Song, Zheng, Chen, Cai, Cao, and Wei, diverted the Yellow River from Xingyang and travelled southeast to meet with the Jihe, Ruhe, Huaihe, and Sihe rivers. The canal connected the Han River and Yunmeng Lake in the west of Chu State, and another was dug to connect the Huaihe River and the Yangtze River in the east. A canal was built to connect lakes and rivers in the Wu State. In Qi State, a canal was dug between the Jihe and Zihe rivers. In Shu State, Li Bing constructed the Dujiangyan Dam to avoid the disaster of a flooded Mohe River; at the same time, two trunk canals were constructed that crossed the city of Chengdu. These canals were all open to boats and could irrigate the surrounding areas with redundant water. This was a definite advantage to the people living nearby.

There were far more canals dug during the Warring States Period than just those recorded by Sima Qian. Nevertheless, Sima Qian aptly described the strategic value of these major artificial waterways. They would eventually evolve into the major sections of the Grand Canal; Honggou became the Tongji Canal, while Hangou was the

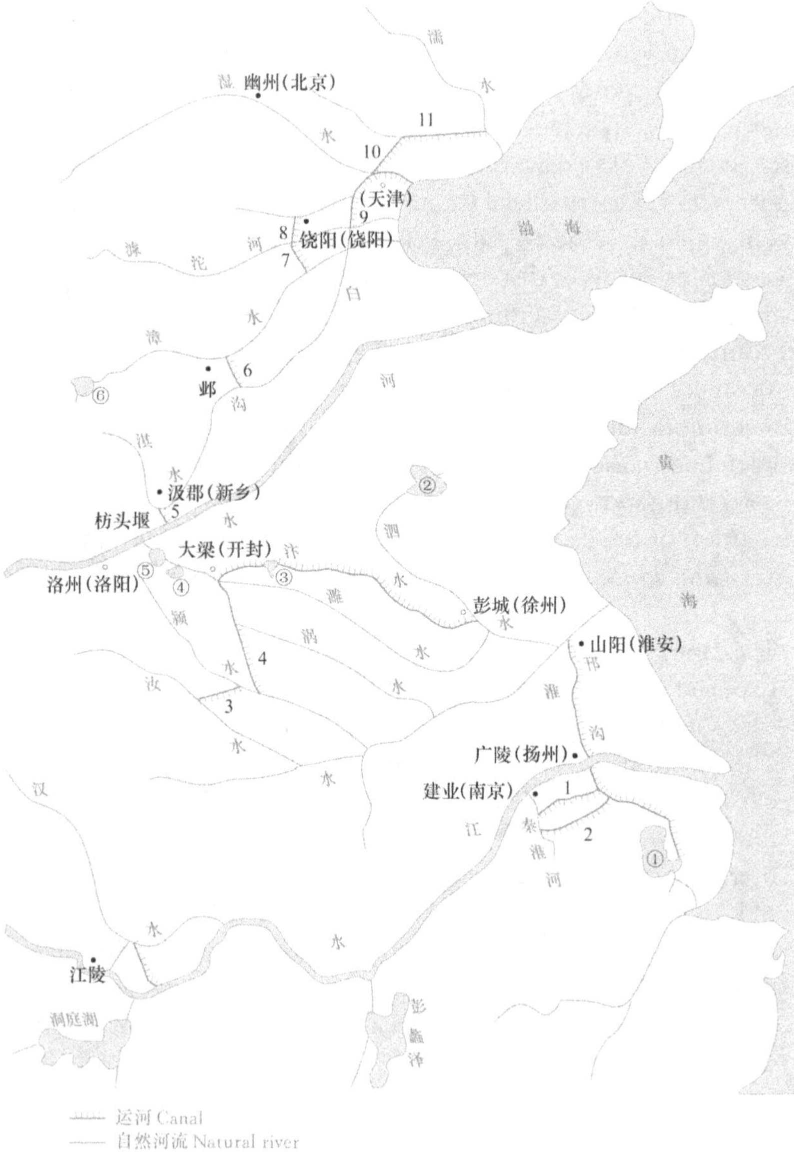

Figure 1.1: The Distribution of Canals in China before the 6th Century

Notes: 1–10: Canals excavated from the period of Three Kingdoms to the Southern and Northern dynasties.

Other canals were excavated from the Spring and Autumn Period to the Warring States Period.

①–⑥: Water tanks or lakes along the canals.

① — Lake Taihu; ② — Juye Lake (Daye Lake); ③ — Mengzhu Lake; ④ — Putian Lake; ⑤ — Yingze Lake; ⑥ — Dalu Lake.

predecessor of the Huaiyang Canal. Others became water-conservancy works that offered many benefits, such as the Dujiangyan Dam. The water transportation through Dujiangyan Dam in Chengdu Plain continued until the 1950s, with the benefits of irrigation lasting until the present day.

1.3.1.1 *The Earliest Artificial Waterways Downstream from the Yangtze River and Around Lake Taihu*

Of the canals recorded by Sima Qian, the canal 'built to connect lakes and rivers in the Wu State' was considered to be the first interval canal. It was formed from numerous artificial waterways excavated from the water-network area around Lake Taihu during the Spring and Autumn Period. In the Three Kingdoms Period, dams and water tanks appeared in the northern intersection with the Yangtze River in Dong Wu (Eastern Wu). This was the initial form of the Jiangnan Canal. During the Tang Dynasty, the Wujiangtang levee was constructed, the canal became separated from the Wusong River and Lake Taihu, and the Jiangnan Canal became an independent engineering system.

During the Spring and Autumn Period, in the powerful state of Chu, numerous artificial waterways were dug in the Jianghan Plain's water network area (the so-called *Yunmeng Ze*, i.e. Yunmeng Marsh, literally 'Great Marsh of Cloud Dream'). The most famous of these was the canal built during the reign of King Ling of Chu (540–529 BC) to connect the Yangshui and Xiashui rivers and achieve the first example of water transportation between plains. The *Commentary on the Water Classic*,[1] written in the 6th century, records this region by stating: 'The canal crossed nearby Zhanghuatai Palace by flowing northward into the Yang River'.

Near Daye Ze (Daye Lake), on the Central Plain, a short canal (Heshui) was dug for the purpose of war. Construction began in

[1] The *Shui Jing Zhu* or *Commentary on the Water Classic* is a work on the ancient geography of China, describing the traditional understanding of its waterways and ancient canals. It was compiled by Li Daoyuan during the Northern Wei Dynasty (AD 386–534). The book is divided into sections according to the different rivers, each of which are described with their source, course, and major tributaries, as well as cultural and historical notes.

482 BC, the 14th year of the reign of King Fuchai of Wu, on the occasion of the meeting of the Central Plain Alliance in Huang Chi (southwest of Fengqiu, in Henan Province). Using the water source of Daye Ze, the canal connected the tributary of the Sishui River (the Qisihi River) and the tributary of the Yellow River (Jishui River). In *Discourses of the States* (*Guoyu*), the author states: '[Fuchai] dug the canal before warring northward. The canal connected Qishui River at the north end and Jishui River at the west end in the area of Shang and Lu states'.

However, King Goujian of Yue, occupying the east of Wu, took advantage of Fuchai's northern expedition and took over the capital of Wu (now the city of Suzhou). King Fuchai of Wu had to lead his army back home, and the state of Wu was destroyed 4 years later. The Heshui Canal was connected with the Jishui River in Dingtao, Shandong, and with the Qishui River in Yutai. After the death of the Wu State, Heshui was abandoned.

1.3.1.2 *Canals Crossing Mountains in the Mid and Low Reaches of the Yangtze River: The Xuxi and Pogangdu Canals*

Of the early canals, the most outstanding engineering and technical achievement was the canal linking the Yangtze River and Lake Taihu. Today, to the south of Nanjing, the Mao Shan mountain range (from north to south) lies between the basins of the Yangtze River and Lake Taihu. During the Spring and Autumn Period, there were two canals across the Mao Shan from the north and south. The southern canal was the Xuxi Canal, which connected a tributary of Lake Taihu (Jingxi) in the east, and connected tributaries of the Yangtze River (the Qingyi and Shuiyang rivers) in the west. The northern canal was the Pogangdu Canal, which started from Jurongxiaoqi and connected with the QinHuai River in the west, linking with the southeast canal in the east. These two canals opened the waterways between the east of the Yangtze River and the basin of Lake Taihu (see Figure 1.2).

The Xuxi Canal was located in the hill region of Mao Shan — the watershed between Lake Taihu and the Yangtze River. The highest elevation of Mao Shan was approximately 20 m and the width of the

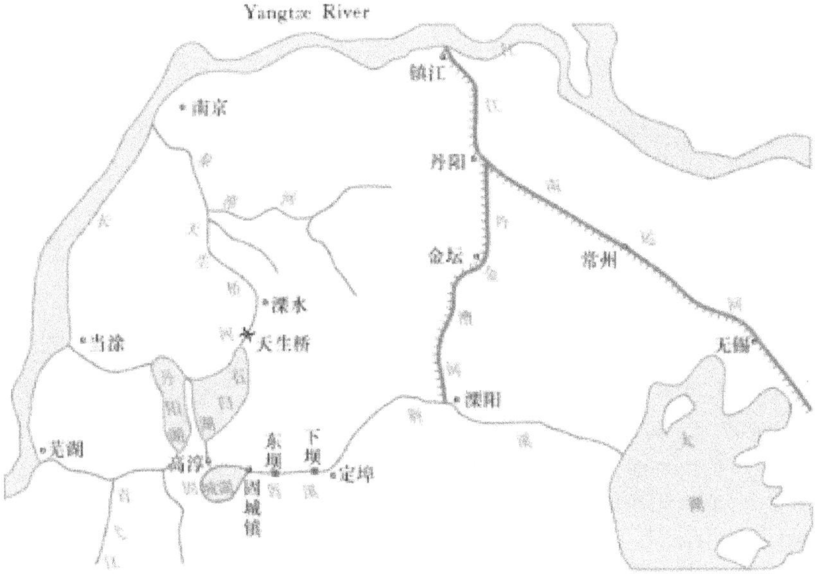

Figure 1.2: The Connections between Xuxi Canal, the Yangtze River, and the Basin of Lake Taihu

hills was 7.5 km along the east–west direction. The elevations to the west and east decreased to approximately 8 and 6 m, respectively. The main engineering facilities of the Xuxi and Pogangdu canals were the weir dams arranged along the hills. Dams were built on different levels to collect surface water and eventually form the waterway. The height of a dam was determined based on the elevation difference between the topography and water levels. The dams formed multi-level waterways. Boats crossed the dams by being pulled along either by humans or animals, making transportation along the water possible in the hilly areas.

King Helyu of Wu (who reigned from 514 to 496 BC) attacked the state of Chu in 506 BC, and Wu Zixu suggested that he constructed a canal to transport grain there and back. The so-called 'Xuxi Canal' connected the Lake Taihu tributary (Jingxi) and the Yangtze River tributary (Yangjiang River). In *Records of the Grand Historian: Treatises of Rivers and Canals*, Xuxi is described as one of the canals in the state of Wu, connecting three rivers and five lakes.

During the Northern Song Dynasty, Shan E recorded the following in *Water in Wu*:

> In spring and winter, two hundred dans[2] of goods were transported in the canal. In the east, boats could move to the Great Lake, while in the west they could move to the Yangtze River ... Cattle on the banks were used to pull boats. At the side of Gucheng Lake, there were officers and offices to collect tax.

Before the Northern Song Dynasty, the engineering facilities of the Xuxi Canal were unknown. During the Song Dynasty, there were five weirs — also known as 'Wu Weirs' — along the 7.5-km-long canal. From east to west, the five weirs were Yujia, Hejia, Kuli, Fenshui, and Yinlin, among which Fenshui and Yinlin weirs were the most important. During the Ming and Qing dynasties, Xuxi was also known as 'Dongba Canal'. To the east of Gucheng Lake, Tianshengqiao Canal (now known as the 'Rouge River') was constructed to connect the Xuxi canal with the QinHuai River. The canal connected the waterway between the Yangtze River and Lake Taihu (see Figure 1.2).

1.3.1.3 *The Connections between the Yangtze River and Huaihe River, and the Yellow River and Huaihe River: The Hangou and Honggou Canals*

We know when the canals connecting the Yangtze River and Huaihe River, and the Yellow River and Huaihe River were constructed. The *Zuo Zhuan* (The Commentary of Zuo) records the construction of the famous Hangou Canal in 486 BC, connecting the Yangtze River and the Huaihe River. Han City was in the northwest region of what is now Yangzhou. The fortification and excavation of the canal were the key projects for the state of Wu to allow them to move into the north. The canal was named 'Hangou' after Han City. At that time, Hangou was just various sections of waterways connecting different lakes. This remained the same until the Eastern Han Dynasty, when Hangou underwent a big change.

---

[2] Dan is an ancient Chinese volume unit. 1 dan equals 100 L.

According to the *Commentary on the Water Classic*, during the 6th century the Hangou Canal flowed from the north of Guangling (modern-day Jiangdu) to Wuguang Lake, Luyang Lake, Fanliang Lake (all the lakes in modern-day Gaoyou), Bochi (in modern-day Baoying), and finally to Sheyang (in modern-day Baoying and Huai'an). There were many lakes and rivers between Huai'an and Yangzhou. The Hangou Canal was constructed by the state of Wu within a very short period, and it cleverly connected these bodies of water. At that time, the Hangou route was in an Ω shape. Hangou had adequate water sources and was the south–north channel that connected the Yangtze and Huaihe Rivers. Therefore, as soon as Hangou was built up, it became a main waterway. In order to avoid stormy waves from lakes and to reduce the risk of circuitous journeys, the section from Jinhu Lake to Baima Lake was constructed during the Three Kingdoms Period (AD 220–280). The canal route began to be straightened. The Huaiyang Canal was very straight until the time of the Song Dynasty.

According to *Records of the Grand Historian: Treatises of Rivers and Canals*, Honggou was one of the most influential canals in the Central Plains and the waterway connecting the Yellow River and the Huaihe River. Honggou was built during the middle of the Warring States Period, and it was an important factor in King Hui of Wei attaining control over the Central Plains. In the 9th year of King Hui's reign (362 BC), the capital of Wei was moved to Daliang (present-day Kaifeng, in Henan Province). Honggou was built the following year. During the Han Dynasty, Honggou was also known as 'Langdang Canal', which was a water network connecting the rivers of Ying, Wo, Yingou, and Sha (the present-day Sui River), also known as the 'Honggou Water Network'. Honggou drew its water from the Ying Ze and Putian Ze (west of modern-day Zhengzhou, Henan Province), which were connected with the Yellow River. From the southeast, Honggou crossed Chenguo (modern-day Huaiyang, Henan Province) and connected with the Huaihe River southwest of the Shou County, Anhui Province.

The early technical aspects of Honggou in water use and connection between artificial waterways and natural rivers were similar to those of Hangou. The water in the Honggou Canal was dependent on the Yellow River. During the Warring States Period, due to the sedimentation settling in Ying Ze, Putian Ze, and the other surrounding lakes, there was no serious problem of silting in Honggou. As a result, during the Sui Dynasty, the Tonji Canal (or the Bian River) was constructed to use the waterways of the 'Honggou Water Network'. The Ying Ze and Putian Ze lakes interconnecting with the Yellow River were filled with sedimentation, which became the toughest engineering problem for the Tongji Canal.

1.3.1.4 *The Connection between the Xiangjiang and Lijiang Rivers: The Lingqu Canal*

After the First Emperor of China, Qin Shi Huang, unified the Central Plains, his army immediately swept Lingnan (the southern region of the Five Ridges). The Lingqu Canal was constructed to guarantee military supplies and control over Lingnan. The canal crossed the watershed between the Xiangjiang and Guijiang rivers, overcoming a terrain elevation difference of 6 m. No canal crossed the watershed with a higher elevation difference until the Huitong River was constructed during the Yuan Dynasty. Due to successful planning, a steady water source and basic navigation conditions were available even during the early stages of the Lingqu Canal's construction. Simple engineering facilities were used, laying the foundations to improve the canal later. Lingqu became an important traffic channel between the Central Plains and Guangdong Province, as well as the Guangxi Zhuang Autonomous Region. During the Qing Dynasty, the number of ships passing through the canal was over 200. In 1941, after the construction of the railway between Hunan and Guangxi, the Lingqu Canal gradually became a channel that was mainly used for irrigation.

There was a geographical corridor (the Xianggui Corridor) between Yuecheng Mountain, in the northwest of Xing'an County, and Dupang Mountain in the southwest, in the Guangxi Zhuang Autonomous Region. The corridor was the overland route from Lingnan to the Central Plains. Yuecheng Mountain was the watershed between the Yangtze and Pearl Rivers, with a ridge height of 20 m and a length from north to south of 300 m. The Xiangjiang River flowed from the south to north and then went to the northeast. The Shi'an River (a tributary of the Lijiang River) flowed from the north to southwest. The nearest distance between these two rivers was only 1.6 km. However, the elevation of the Xiangjiang River here was 6 m below that of the Shi'an River. Therefore, the Lingqu route was not designed here by connecting the two rivers linearly. Upstream from the Xiangjiang River, engineers in the Qin Dynasty looked for the water-division point connecting the canal and the two rivers. South and north canals were constructed; the north canal was connected with the Haiyang River (a tributary of the Xiangjiang River) and the south canal was connected with the Shi'an River.

The longitudinal slopes of the south and north canals were reduced through curve routes so that the artificial channels and natural rivers were smoothly connected, thus giving suitable water velocity and depth for navigation. According to a survey carried out in the 1940s, if the north canal was arranged in a straight line, the slope was approximately 3.75 per cent; with a curve length of 3.25 km, the average slope was reduced to 1.7 per cent. As for the south canal, with a curve length of 33.15 km, the average slope was only 0.9 per cent. With the reduced slopes and the help of manmade power, the Lingqu Canal provided the most basic navigation conditions.

The development of engineering technology in the Lingqu Canal reached its peak during the Tang Dynasty. Its basic structure was later inherited. During the 1st year of the Baoli reign (AD 825), Li Bo renovated the Lingqu Canal. At the division point, Hua Dike (later called 'Spade Mouth') was designed for water diversion. On the canal, 'doumens' (or simple gates) were set up to control the water depth. The navigation conditions were far improved compared with the Qin Dynasty. Detailed descriptions of the Hua Dike and document of Lingqu emerged during the Song Dynasty. The shape and structure of the Lingqu Canal headwords has not changed significantly since then.

1.3.2 *From the Three Kingdoms to the Northern and Southern Dynasties: Preliminary Formation of the National Water-Transportation Network*

A new era for the canals began during the Three Kingdoms Period and the Jin, Northern, and Southern dynasties. A new canal network began to form to connect the Yangtze River, the Huaihe River, the

Yellow River, and the Haihe River. To the east of the Qiantang River, the Zhedong Canal formed the basic framework through the natural improvement of the river and the construction of weir dams. All these achievements laid the foundation for the Sui and Tang Grand Canal.

1.3.2.1 *Canals Connecting the Haihe River Basin*

The Haihe River basin was divided into the north and south basins by the Juma River. The rivers of the south basin were tributaries of the Yellow River, but the construction of the canals changed this distribution. The southern and northern canals formed the main streams of the south and north basins, respectively. During the late Eastern Han Dynasty, Cao Cao (AD 155–220) began military operations against Yuan Shao to unify the north. In the 9th year of Jian'an's reign (AD 204), Cao Cao ordered the construction of the Fang Dam to lead the Qishui River into the Baigou Canal, which was used to transport grains. The Qishui River originated from Taihang Mountain and entered into the Yellow River from north to south. The Qingshui River (upstream of the Weihe River) joined the Qishui from the southwest. The old course of the Yellow River from the Han and Tang dynasties was utilised in the construction of the Baigou and Qishui canals. In the 18th year of Jian'an's reign (AD 213), the Licao Canal was constructed near Yecheng to lead the Zhangshui River into the Baigou. The waterway therefore travelled from Yecheng to the Yellow River via the Licao and Baigou canals, then on to the Yangtze and Huaihe Rivers.

During the 11th year of Jian'an's reign (AD 206), Cao fought with Wuhuan in the north. In order to facilitate transportation to the north, the Pinglu, Lukou, and Quanzhou canals were built, connecting the White River of the Haihe River basin with the Zhangshui, Hutuohe, and Chao rivers. These three waterways were equivalent to the current South Canal and North Canal. In the 4th year of Emperor Yang of Sui's reign (AD 608), the Yongji Canal was constructed along the old course of the Baigou. The southern and northern main stream systems of the Haihe River basin were then finally established.

1.3.2.2 *The Renovation of the Sishui and Bianshui Waterways*

In May of the 6th year of Liang Putong's reign (AD 525), the construction of the Suyu Weir connected the waterway from the Huaihe River to the Sishui River. Suyu was located in Tonghuai, downstream from the Sishui and in the southeast of Suqian County. The famous Sishui rapids, Xuzhouhong and Lyulianghong, were in the southeast of Pengcheng (present-day Xuzhou), approximately 25 km away from the centre. The renovation of Xuzhouhong began during the Eastern Jin Dynasty. This waterway became a section of the Tongji Canal during the Sui Dynasty. In the 12th year of Yonghe's reign during the Eastern Jin Dynasty (AD 356), the Sishui was connected to the Wenshui and Jishui rivers from the north of Pengcheng, reaching Dong Ah in the north. This was the predecessor of the Jizhou River in the Yuan Dynasty.

During the Cao Wei period, the canal connected the Huaihe River with the Sishui River, the Bianshui River, and the Yellow River in the west. In the 7th year of Jian'an's reign (AD 202), the Juyang Canal was built from Junyi (modern-day Kaifeng) to Juyang (in the south of modern-day Shangqiu). It formed one section of the Bian Canal during the Tang and Song dynasties. In the capital of Northern Wei (modern-day Luoyang), the Bianshui was dredged from AD 500–508, around 100 years before the construction of the Jiqu Canal during the Sui Dynasty.

1.3.3 *The Sui, Tang, and Song Dynasties: The Formation of the Grand Canal Engineering System*

After the collapse of the Qin–Han unified empires and 400 years of separatist wars during the Wei and Jin dynasties, China once again entered a period of unity in the Sui, Tang, and Song dynasties. A waterway linking the political and economic centre became the foundation of the country. During the Sui Dynasty, the Grand Canal was renovated and built up, with connections made from east to west and north to south. The Tang and Song dynasties were the second historical climax of canal construction and a brilliant period of innovation. In terms of engineering, the systematic renovation of rivers meant that water-source projects (such as water tanks and lakes) and water control projects (such as gates and dams) became independent engineering systems in the Grand Canal. The water-transportation system also reached its historical peak during this period; through the combination of warehousing and waterways, water transportation was well-organised and highly effective. In the later dynasties, the canal was operated in the same way.

The Tongji and Yongji canals were constructed within a very short period during the reign of Emperor Yang of Sui (AD 569–618), which was undoubtedly a result of the success of the previous generations. The Tang and Song dynasties were the period during which the final engineering system of the Grand Canal was established, with the highest scientific and technological achievements being reached during the Song Dynasty. During the Northern Song Dynasty, the Guishan Canal was constructed at the northern end of the Huaiyang Canal. The Guishan Canal crossed the Huaihe River in Guishan town and entered into the Bianhe River in the city of Sizhou. At the north and south joints of the canal and the Huaihe River, dams and gates were used to control the water so the two rivers were smoothly connected and vessels crossed the rivers easily. The construction of the Guishan Canal marked the clear boundaries between the artificial waterways and natural rivers.

The Huaiyang Canal became a relatively independent engineering system. During the Tang and Song dynasties, perfect water control projects (doumen) and water-supply projects (water tanks) were built along the canals. Simply put, the canals were separated from the natural rivers; water entered the canals through the joints and key locations (watersheds). From the Song Dynasty onwards, the word 'canal' referred specifically to artificial waterways. During this period, besides the Bianhe Canal, the artificial waterways to the south of Huaihe River were mostly called 'canals', such as the Huaiyang Canal (or Huainan Canal), Gaoyou Canal, Yangzhou Canal, and Lin'an Canal.

During the Song Dynasty (AD 960–1279), a complex gate with comprehensive engineering effectiveness was constructed in the Huaiyang and Jiangnan canals. The facility was a well-designed and key water control system that consisted of a gate, diversion and withdrawal drains, and an *Ao* (water storage pond). Through the clever arrangement of the gate, the *Ao*, and drains, and the synchronisation of the gate's opening and closing with the rise and fall of the tide, the facility was able to function as a method for tide control, water storage, water conservation, and water delivery. It was also used as a waterway lock, similar to modern gates. The ship lock in Milan during the 17th century was almost identical to the Zhenzhou gate of the Song Dynasty. These complex gates were used for approximately 100 years, until the Yuan Dynasty, when they were almost entirely replaced by weir dams. During the Yuan, Ming, and Qin dynasties, at the joints of Beijing–Hangzhou Grand Canal, weir dams were used to store water for transportation.

The Grand Canal had been extended to Ningbo during the Tang and Song dynasties. The waterways to the east of Zhejiang were basically canalised. The canal became an important inland waterway, connecting with the open sea and the waterway from Ningbo to Bianzhou. Over the course of the project, the Jiangnan Canal was naturally connected to the natural rivers. However, the Zhedong Canal was a canalised waterway. Through the use of plate dams, vessels could sail against the current and move smoothly between the canal and the rivers.

The major technical achievement of the Grand Canal during the Sui, Tang, and Song dynasties was the establishment of a continuous transport system from the Zhedong Canal to the Bianhe River. The canal sections in the north of the Qiantang River became a relatively independent engineering system. The continuous transportation waterway stretched for over 2,000 km from north to south. At a joint of a canal and a natural river, the canal and the river could be smoothly connected through engineering control. The joints performed the integrated functions of traffic control, water supply, sediment control, and so on. The boundaries between artificial waterways and natural rivers became increasingly clear. The complex gates at the joints between the canal and the Yangtze River and the Huaihe River were the landmark projects, pioneering the world's hydraulic engineering techniques during this period.

1.3.4 *The Yuan, Ming, and Qing Dynasties: The Beijing–Hangzhou Grand Canal and its Technical Achievements*

The construction of the waterway from the south to Beijing began immediately after the city became the capital during the Yuan Dynasty. The Tonghui and Huitong canals built in the late 13th century connected with the northern section of the Yongji Canal (the Baihe Canal, now known as the North Canal), the Wei River (now known as the South Canal), and the Huaiyang Canal, the Jiangnan Canal, and the Zhedong Canal to the south of the Yellow River. This formed the north–south water channel, with Beijing and Hangzhou as the terminals — the modern-day Beijing–Hangzhou Grand Canal. During the Ming and Qing dynasties, the canals operated by the state for water transportation were called 'Caohe'. The 'Caohe' managed by the state stretched from Beijing, in the north, to Yangzhou, in the south. The River Governor and Transportation Governor were the highest chief executives to manage the engineering and transportation across the canal, respectively. The Jiangnan Canal, to the south of the Yangtze River, and the Zhedong Canal belonged to the local water-conservancy project, which was managed and maintained by the local government.

During the Yuan, Ming, and Qing dynasties, with Beijing as the terminus for water transportation, the Beijing–Hangzhou Grand Canal connected to five large rivers. The length of the continuous waterway was 1,700 km. The differences in regional water-resource conditions and terrain elevations along the canal were far greater than those in any previous canal system. Furthermore, the operation of the Beijing–Hangzhou Grand Canal was during the period when the Yellow River crossed the Huaihe River and flowed to the south. The massive amount of sediment carried by the flood from the Yellow River greatly interfered with the canal; the scale and difficulty of the water-control activities needed to ensure unimpeded travel along the canal were unprecedented. The operation of the Beijing–Hangzhou Grand Canal lasted up to 500 years. Hundreds of kilometres of levees and numerous dams and spillways were built along the banks of the

Yellow River. Joints were built throughout different periods along the section of the Yellow River from Xuzhou to Suqian. At the intersection of the Yellow River and the Huaihe River, the joint Qingkou–Hongze Lake project lasted for more than 300 years.

1.3.4.1 *The Huitong River: The Canal Crossing the Largest Terrain Elevation Difference before the 13th Century*

During the Yuan Dynasty, there were two routes from the south to the capital (Beijing). The first was by sea to Tianjin, then over land to Beijing. The second was to travel along the Yellow River to Fengqiu, Henan Province, then over land to Qimen, along the Yuhe River to Tongzhou, and then over land to Beijing. However, the Yuan Dynasty saw the construction of the Tonghui Canal (from Beijing to Tongzhou) and the Huitong Canal (from Linqing, Shandong, to Xuzhou). The Yuhe and Sishui rivers were connected by a waterway, and thus the Grand Canal ran all the way from Beijing to Hangzhou.

Careful planning was needed in the construction of the Tonghui and Huitong canals. The preliminary work included surveys of a wide range of water resources and land. During the Yuan Dynasty, the water monitor Guo Shoujing (AD 1231–1316) was a famous astronomer and water expert. A contemporary record describes him measuring the headwater of the Wei River, downstream of the Yellow River from Mengmen, and comparing the altitude of the land at Beijing and Bianling. He led this survey, which offered a wide range of altitude data and other important information along hundreds of miles of the old course of the Yellow River. This data provided the basis for the planning of canal routes in the north of the Yellow River.

The construction of the Jizhou Canal began in the 13th year of the Yuan Dynasty (AD 1276), a 65-km-long new section being built from Jizhou to Anshan (modern-day Dongping County, Shangdong). The canal connected with the Sishui waterway (which was constructed during the Jin Dynasty) in the south and the Daqing River, Dong Ah, in the north. The Jizhou Canal's water was drawn from the

Wenhe and Sihe rivers. The canal was finished by the 20th year of the Yuan Dynasty (AD 1283). Water transportation began from Lyuliang, Xuzhou, the north shore of the Yellow River, crossing Shandong and continuing on to the north.

The operation of the Jizhou Canal demonstrated the feasibility of inter-basin transfer of water resources, which was a crucial step towards the ultimate connection of the canal with the Yuhe, Wenhe, and Sihe rivers and the smooth crossing of the areas with poor water resources. After the construction of the Jizhou Canal, one section between the Yuhe and Sihe rivers was left unconnected. In the 26th year of the Yuan Dynasty (AD 1289), under the direction of official Ma Zhizhen, the construction of a canal began from the south of Xucheng, Ansan, to Dong Ah, Liaocheng, and Linqing, stretching for more than 125 km. Kublai Khan named the canal the 'Huitong River'; later, the Huitong and Jizhou canals together shared the name of 'Huitong River'.

The Huitong Canal crossed the Shandong host with its highest elevation point at Nanwang, in northern Jining. During the Yuan Dynasty, the Huitong water-diversion project was the Huiyuan Gate (known as the 'Tianjin Gate' in the Ming and Qing dynasties) in Jining. The Huiyuan Gate was 40 km away from the highest elevation point, Nanwang, with a difference in elevation between the two of approximately 10 m; as a result, it was difficult to supply water to the Nanwang section. During the Yuan Dynasty, transport along the water of the Huitong Canal was not smooth, and the annual amount of grain carried along the canal was less than 200,000 dans (1 dan is equal to around 60 kg). In the 9th year of the reign of the Yongle Emperor, the head of the Ministry of Works, Song Li, oversaw the movement of the water project northwards to Nanwang, Wenshui. The project was introduced in the southwest, entering the canal at Erhekou, Wenshang County. From then onwards, the water flowed northward at a speed of 4/10 and southward at a speed of 6/10 from Nanwang, which was the highest portion of the section. Sluices were built in the proper sites to control water flow in the canal. The water-shortage problem in the uphill section of the canal, in the north of Jining, was finally resolved.

From the 9th year of the reign of the Yongle Emperor (AD 1411) to the 21st year of the reign of the Wanli Emperor (AD 1593), the Daicun Dam and Gangcheng Gate were built and the Wenshui River was introduced to the south to fill water tanks in Nanwang. The Daicun Dam was originally a temporary rock-filled dam, but it was changed to a permanent masonry weir during the 17th year of the Ming Dynasty (AD 1589). In the Qing Dynasty, the Daicun Dam became a more successful river dam, with a length of 437 m. According to its different structures and functions, the whole dam was divided into an overflow dam, a rubble dam, and an exquisite dam. The overflow dam was a masonry dam. Located in the middle, the rubble dam was a rock-filled dam with sand seepage and water seepage, designed to separate the sediment from the water. The exquisite dam was made of permeable stone and performed various functions, including damming the water, diversion, desilting, and providing a spillway.

With the concurrent operation of the Daicun Dam, the key water-control project in Nanwang was used for water regulation and diversion to the north and south. After the completion of the Daicun Dam, the Xiaowenhe River drew its water from the lakes of Nanwang. Water tanks in Nanwang provided sufficient water for the highest section of the north and south divisions of the Huitong Canal. The Daicun Dam and Nangwang water control project were continuously improved. During the Ming Dynasty (around the late 15th century), Wang Qiong recorded in the *Caohe Pictorial: Caohe* that the north–south water diversion in the Nanwang watershed had achieved qualitative control, after which the water flowed northwards at a speed of 4/10 and southward at a speed of 6/10 from Nanwang, which was the highest portion of this section. Sluices were built in the proper sites to control water flow in the canal. The north–south water diversion was achieved using the north and south gates — later known as Shili Gate and Liulin Gate, respectively. The lakes in Nanwang were used as water tanks to store water as well as sediment. With the addition of the water and sediment tanks, dredging was concentrated on these instead of the whole canal. The amount of time between dredges was extended by a few years.

1.3.4.2 *An Outstanding Achievement in Water-Resource Planning: The Tonghui Canal Water Project*

Beijing is located in the alluvial plains of the Yongding and Chaobai rivers. There was no natural water supply in the region to supply the canal, and the only feasible way of overcoming this difficulty was to transfer water from far away. Beijing was designated as the capital during the Jin Dynasty, and plans to construct the canal were set in motion. The water for the Yongding River was drawn from Shijingshan, in southwest Beijing, into the Jinkou River, entering into the Caohe River at the southeast through the Sanlihe River. The Yongding River had a high concentration of sediment, and its water level changed greatly between the flooding and dry seasons. The Jinkou River was filled with sediment in the Yuan Dynasty.

During the Zhongtong Era (AD 1260–1264), the Imperial astronomer Guo Shoujing brought six water-conservancy projects to the attention of Kublai Khan, two of which were related to water transportation to the capital. The first was the digging of a canal to introduce water from the Yuquan River into Tongzhou, in the east. The second was the dredging of the Baihe and Yuhe rivers, south of Tongzhou, to restore water transportation from Tongzhou to Linqing. In the 28th year of the Zhiyuan Era (AD 129), Gou Shoujing created a comprehensive plan for the canal's water sources, water regulation, and water storage; he proposed the diversion of water from Shenshan Spring (in the village of Baifu, Changping County) westward and then south, taking in the Shuangta, Yuhe, Yimu, Yuquan springs and others. The waterway would then flow into the capital, across the west gate, and converge into Jishuitan Lake in the south before crossing Wenming Gate (the east gate) and flowing into the Baihe River. The waterway would have a length of 164 li[3] and 104 bu[4] and 12 sections of dike, with a length of 310 bu, along the channel to block the inflowing streams. Twenty sluices would be built at 10 locations to control water level and flow.

---

[3] 1 L equals 500 m.

[4] 1 bu equals to the distance covered by a step.

Guo Shoujing was put in charge of the construction of the canal in the 29th year of the Zhiyuan Era (AD 1292), and his plans were fully implemented. The water-source project for the canal was constructed upstream of Shichahai, including facilities for drawing springs, water transfer, and water storage and regulation. Through water-conservation measures, spring water from the Beijing Xishan was introduced into the Wengshanbo Lake (now Kunming Lake). Water was then transferred into ponds and lakes in the city through channels, forming different lakes across the city, such as today's Beihai and Zhongnanhai lakes. The lakes reregulated the water and supplied it to the canal in a controlled manner. The elevation difference between Beijing and Tongzhou was approximately 20 m; 20 consecutive gates were built along the Tonghui Canal to regulate water and guarantee a depth that would be navigable for ships. The municipal and environment functions of the Tonghui Canal and the water-source works became increasingly important; the water projects of the canal gradually evolved to form the modern river and lake system in Beijing.

1.3.4.3 *Qingkou: The Key Water-Control Project of Hongze Lake*

From the 2nd year of the Jinyan Era during the Southern Song Dynasty (AD 1128), the Yellow River changed its course to move south. After 400 years of use, the section of the river from the estuary to Xuzhou was beset with a serious silt problem. The canals connected to the Yellow River were affected by the bursting of the river, which happened frequently from the mid-Ming Dynasty onwards. The bursts and changes of flow paths resulted in dry canals, and boats were unable to cross the Yellow River, becoming blocked south of Huai'an.

In the 6th year of the reign of the Wanli Emperor (AD 1578), the Governor for Rivers and Canals, Pan Jixun (1521–1595) was placed in charge of constructing the Gaojiyan Dam. Pan Jixun proposed and applied a method of controlling sedimentation by flushing it with water that had been stored. Taking into account the characteristics of the upstream and downstream rivers of Qingkou, he led the planning and construction of the following projects: the guided works on the Yellow River, the Huaihe River, and the canals; the water-storage project on the Huaihe River; and the dike project on the Yellow River, the Huaihe River, and the canals. The key water-control project on the Qingkou–Hongze Lake was thus an engineering system with multiple functions.

The Gaojiyan Dam was the key work of the Qingkou–Hongze Lake water-control project. The dam extended south from Qingkou for a length of approximately 60 km, an average height of 12 m, and a maximum height of 19 m, resulting in the formation of the Hongze Lake. Within the dam, sections were constructed for backwater and flood reduction. The clean water of the Huaihe River was stored within the backwater dam and then introduced from Qingkou to flush sediment out of the Yellow River. Floodwater flowed over the flood reduction dam into the Gaoyou and Baoying lakes. From the construction of the Gaojiayan Dam during the Ming Dynasty and after continued construction during the late Ming and early Qing dynasties, the Qingkou–Hongze Lake water project became a perfect water-control project, featuring all the basic engineering properties found in modern reservoirs (see Figure 1.3).

In the 26th year of the reign of the Kangxi Emperor (AD 1686), the Zhong Canal (later known as the Middle Canal) was constructed to the north of the Yellow River, beginning in Luomahu and going to Suqian, Taoyuan, and Zhongjiazhuang, Qinghe County. Grain boats from the south were now able to go north to Qingkou, through the Yellow River for a few kilometres, and then, finally, to the Zhong Canal. In this way, boats could avoid going along the Yellow River for 90 km, and the canal and the Yellow River were separated. The formation of Hongze Lake meant it became the water source for the northern portion of the Huaiyang Canal.

The artificial lake, Hongze Lake, was created through changes in the natural environment while the Yellow River crossed the Huaihe River and after 400 years of large-scale water-conservancy and canal maintenance activities. The water-storage capacity of the lake reached 3.5 billion m^3. The Qingkou–Hongze Lake water project achieved seasonal water adjustments in the Huaihe River and controlled the

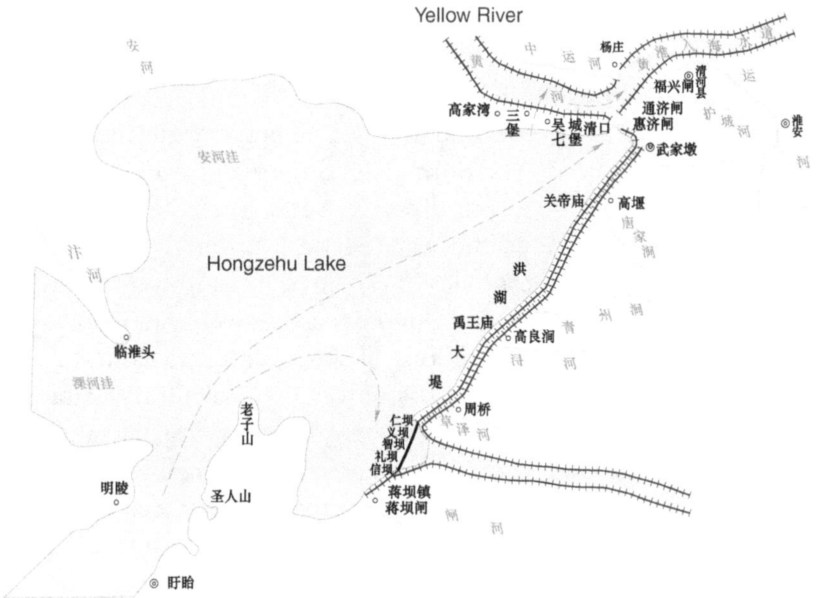

Figure 1.3: The Qingkou–Hongze Lake Water Project

amount of water and water levels in the canal at Qingkou, connecting the Yellow River and Huaihe River smoothly. After the Huaihe River changed its path into the sea, Hongze Lake became an important water source for the northern Jiangsu region. In the near future, Hongze Lake will be one of the key storage reservoirs for the large project to divert water from the south to the north.

The complete engineering system of the Hongze Lake represented the world's highest level of civil engineering technology before the Industrial Revolution in the West in the 17th century. The Gaojiayan Dam was the largest dam before the 17th century and a milestone in the history of dam constructions.

1.3.4.4 *From the Canalised Waterway to the Engineering System of the Zhedong Canal*

The Zhedong Canal was the eastern extension of the Beijing–Hangzhou Grand Canal. From the north, the canal connected with the Qiantang River at Xinxitang, Xiaoshan District, and reached the mouth of the Yongjing River at Ningbo, in the south, and approximately 200 km across the Xioshao Plain.

The construction of the Zhedong Canal began from the Shanyin waterway in the state of Yue during the Spring and Autumn Period. From the year 457 to 464, spur dikes were constructed using bamboo or wooden cages to guide the water. Later, Nanjin Dam was built here, one of the earliest examples of canalised natural rivers in eastern Zhejiang. During the Tang and Song dynasties, weirs and dams were constructed at the key intersections of waterways and natural rivers, essentially controlling the depth of the water to make it navigable. At that time, the canal crossed three rivers (the Qiantang, Puyang, and Cao'e rivers) from Hangzhou to Ningbo and then finally turned into the Yuyao River and the sea. The canal crossed seven weirs from the west to the east — namely Xixing (the joint with the Qiantang River), Qianqing (the joint with the Puyang River), Dusi (the inlet of Jianhe Lake), Cao'e (the joint with the Cao'e River), and Lianghu, Tongming, and Xidu (the dams on the Yaojiang River). By running the weirs (or gates) at the joints, the Zhedong Canal introduced water for transportation and overcame the differences between the canal and the natural rivers.

From the Northern and Southern dynasties to the Ming Dynasty, the Zhedong Canal remained the only form of a canalised waterway. In the 16th century, the Sanjiang Gate was built, and the section from Shaoxing to Xiaoshan — which was impacted the most by natural rivers — became an independent engineering system. This was the major turning point in terms of engineering technology along the canal.

The Xixing Canal, north of the Cao'e River, was a section of the Zhedong Canal from Xiaoshan to Shaoxing. The canal was connected with the Puyang River, which originated from Tianmu Mountain. The canal moved eastwards through the joint of the Qianqing weir in the Puyang River, entering into the Qiantang River in the north of Shaoxing. The Puyang River was short, flowed rapidly, and was subject to the effects of tidal backwater. The northern section of the Zhedong canal was therefore greatly influenced by the natural river.

In the middle of the Ming Dynasty (*c.* 1530s), the Puyang River diverted to the northwest and entered into the Qiangtang River from Yupu, Xiaoshan District, no longer crossing the canal. The conditions of the canal were thereby improved, but water supply became a prominent issue. In the 16th year of the reign of the Jiajing Emperor (AD 1537), a sluice (Sanjiang Gate) was built at the joint of the Puyang River (or the Xixiao River), Shaoxing County, the Cao'e River, and the Qiantang River. The Sanjiang Gate had a total length of 103 m and 28 gates. To tackle the low water levels, the gate was opened to bring water from the Cao'e River into the canal; during typhoons, the gates were closed to block saltwater intrusion from the tides and to store fresh water. After the Sanjiang Gate had been completed, the amount of water in the Zhedong Canal from the north (the Xixing Canal) to the south (the city of Shaoxing) could be regulated. The section of the Zhedong Canal from the Qiantang River to the Cao'e River became an independent engineering system. The following section of the Zhedong Canal was connected to the canalised natural waterways of the Yuyao River and finally entered the sea through Sanjiangkou, Ningbo.

The utilisation of the Sanjiang Gate meant that quantitative control of the canal water was achieved, which also provided the Xiaoshao Plain with the comprehensive benefits of irrigation, urban water supply, flood control, and drainage. There were strict rules to regulate the opening and closing of the gate, and these rules were printed in front of the gate and on the canal bank outside Shaoxing. The five rules were represented by the characters for gold (金), wood (木), water (水), fire (火), and soil (土). The gold character meant that all 28 gates were opened; the wood character meant 16 gates were opened; the water character meant eight gates were opened; and the fire character meant all gates were closed. The rules were changed in the 1st year of the reign of the Xianfeng Emperor (AD 1851); now, the wood character meant all the gates were opened, the water character meant 16 gates were opened, the fire character meant eight gates were opened, and the soil character meant all gates were closed. The changes in water levels controlled by the front gates in the Ming and Qing dynasties reflected the effects of the change in sedimentation levels outside the gates.

The Sanjiang Gate was managed by the Sanjiang Inspector. Specific gate workers were hired to operate the gates. The gate was built on bedrock, and the gate block and base were made of stone masonry. After more than 400 years, the structure remains largely intact. Due to the accumulation of sediment in the Cao'e River estuary, a new gate was built in 1979, 2,500 m away from the outside of the original gate.

The Sanjiang Gate was a key project on the Zhedong Canal. Its world-leading hydraulic design and structure meant that it took an important position in the engineering and technical history of the Grand Canal.

1.4 The Engineering and Transportation Management of the Grand Canal: The Case of the Beijing–Hangzhou Grand Canal

The transportation of grains to the capital via the canals (the so-called *Caoyun*; *cao* and *yun* meant canal and transportation, respectively) was a crucial part of the food supply to Beijing, its adjoining areas, and the defenders. Indeed, the transportation of grains along the canals provided the foundation for economic and social stability during many historical periods in China. Since the Han Dynasty, all dynasties had endeavoured to secure the ability to transport grains to the capital over water. Liu Yan established the method of water transportation during the Tang Dynasty, and Pei Yaoqing continued to improve this method to achieve a unified and independent management system using the canal waterways and warehouses. During the Song Dynasty, the water transport operation procedures were inherited from the previous generations. Besides grains, tributes and commodities were also included in the transportation management. During the Yuan, Ming, and Qing dynasties, due to the complex terrain and huge regional differences in hydrology and water resources along the Beijing–Hangzhou Grand Canal, transportation along the canal was much more difficult than before. During the Yuan Dynasty, a monitoring office was set up in Zhangqiu to manage the Huitong River. As the Huitong River did not flow smoothly at this time, water transportation was largely reliant on the ocean instead. A unified management system of canals had not been established by the end of the Yuan Dynasty. In the Ming Dynasty, the canal-based water transport system was established to form a strict canal management system from the centre to the locals. The system meant the separation of rivers and canals and the scaling back of the two systems of canal management and transport management. More legislation and the appointment of more officers meant the management system became further refined, guaranteeing the smooth progress of water transport.

1.4.1 *The River-Management System*

During the reign of the Yongle Emperor during the Ming Dynasty (AD 1403–1424), water transportation gradually changed from being carried out on sea to being carried out on the canals. There was no designated water transportation official during the early Ming Dynasty, but when some serious issues arose on the Yellow River or the Grand Canal, a minister would be appointed to manage the situation and resume his other duties after the problem was resolved. In the 15th year of the reign of the Yongle Emperor (AD 1417), the General Commander of Water Transport was appointed to manage transportation along the rivers and canals. Chen Xuan was the first to assume this position. The position of Prime Minister of Rivers (or the Governor of Rivers and Canals) was established in the 7th year of the reign of the Chenghua Emperor (AD 1471). The Governor was in charge of the management of the Yellow River and the canals. Thus, the management of the canals and the management of transport along them became two different systems. The Governor of Rivers and Canals was a standing position, and assistant ministers or ministers were often appointed to it. Additionally, provisional governors and censors (as well as central censors), Imperial guards, and eunuchs were often dispatched to inspect the canals.

From the 13th year of the reign of the Chenghua Emperor during the Ming Dynasty (AD 1477), three 'Water Divisions' were established along the Grand Canal to the north of the Yangtze River — the North River Division (in Tianjin), the East River Division (in Jining), and the South River Division (in Huai'an). The canal began to be managed according to these branches, with ministers from the central government appointed as officials for each of the Water Divisions. The management of the canals and rivers was therefore shifted from central government to local government. The central agencies included that of the Governor of Rivers and Canals, the Prime Minister of Rivers, and the officials of the Water Divisions. The local authorities included the administrative systems of the provinces, Zhous (ancient administrative divisions), and counties. A paramilitary management system — including 'Wei', 'Suo', and 'Qianhu' — was responsible for flood control and security.

During the reign of the Xuande Emperor during the Ming Dynasty (AD 1426–1435), besides the administrative and military systems, official divisions were set up at the canal and river joints, key projects, vulnerable sections, and locations from where the canals drew water. For example, there was the Linqing Sub-Division and the Nanwang Sub-Division. The officer of the sub-division was known as the Principal (*Zhushi*), such as Principal Lyuliang Hong, Tai'an, Shandong Province. In the Zhous along the canal, officers known as *Tongpan* were appointed to manage the rivers. In counties, officers known as *Zhubo* were appointed to manage the rivers. Local officers were responsible for dredging the rivers, patrolling the canals, and preventing theft and fires along the rivers. Some of these sub-divisions were not permanent but needs-based, such as the Shangdong Ningyang Sub-Division.

The river management system from the Ming Dynasty was basically continued during the Qing Dynasty, but it was simplified. The river and flood administration system was hierarchical (River Governor, Dao, Tin, and Xun) in the same way that the corresponding military security systems were. The River Governor was the chief executive. During the reign of the Yongzheng Emperor (AD 1723–1735), four river governors were appointed across different regions to manage the Yellow River, the Huaihe River, and sections of the Grand Canal. The South River Governor (simply known as *Nanhe*) was located in Qingjiangpu and managed the sections of the Yellow River, the Huaihe River, and the canals in Jiangsu and Anhui provinces. The East River Governor (simply known as *Donghe*) was located in Jining and managed the sections of the Yellow River and canals in Henan and Shandong provinces. The Zhili Governor (simply known as *Beihe* or referred to as the North River Governor) was located in Tianjin and managed the rivers and canals in the Haihe River basin. The full-time sub-management system was called *Dao*. For example, during the Ming Dynasty, the Water Divisions included Tongyong Dao (for the Zhili), Canal Dao (in Shangdong), Xuhuai Dao (in Jiangsu), Huaiyang Dao (for the Huaiyang Canal), Changzhen Dao (for the South Canal), and so on. The military rank corresponding to the Dao was the River Standard Lieutenant. The Ting was equivalent to the local Zhou government, administrative officers included *Tongzhi* and *Tongpang*, and the military post was *Shoubei* (garrison). The Xun was equivalent to the county-level government and administrative officers included *Xiancheng* and *Zhubo*, and the military post was *Ganzong*.

During the Ming and Qing dynasties, labourers directly engaged in the management of the rivers that were organised in units known as *Jia*, which had sub-chiefs known as *Laoren* (or *Zong Jia*) and *Xiao Jia*. The labourers (called *Fu*) were divided into *Qian Fu* (who dredged the rivers and guided boats over shoals), *Zhaba Fu* (who dealt with dams), *Di Fu* (who dealt with dikes), *Liu Fu* (who tracked boats), *Quan Fu* (who dealt with springs), *Tang Fu* (who dealt with ponds), and *Hu Fu* (who dealt with lakes). At ports and hubs, there were also *Laosha Fu* (labourers who collected sand), and *Tiaogang Fu* (porters). The labourers came from different sources, including those in military service, those in servitude, and those who had been recruited and earned a wage. During the early Ming Dynasty, the number of standing labourers reached approximately 47,000. During the Qing Dynasty, the number of labourers was less than 10,000, most of whom were temporary workers.

1.4.2 *The Management System of Water Transportation*

During the Yuan Dynasty, the Water Transport Division (*Caoyunsi*) was established in the capital, Nanjing, and was tasked with managing water transportation. In the 14th year of the Hongwu Emperor's reign during the Ming Dynasty (AD 1381), the Water Transport Division was merged with the River Management Division. The Water Transport Division was established again after the 1st year of the Chenghua Era, with the chief executive (known as the 'Canal Governor') placed in charge of water transportation, including nominating officers, constructing and servicing boats, and managing and supervising the transportation of tribute rice. During the Ming Dynasty, a military post, that of the *Caoyun* Commander, was established to escort vessels travelling over water. Thus, water transportation was managed by two systems of transport and escort. The post of *Caoyun* Commander was abolished during the Qing Dynasty, with the Army used for escorting duties instead.

The Canal Governor and *Caoyun* Commander were responsible for the collection, transportation, and warehousing of grains. They were also responsible for the assessment of river officials and engineering inspections. The two systems of transport and escort were mutually restrained, and friction developed frequently. During the Ming Dynasty, the Canal Governor and *Caoyun* Commander were required to go to Beijing each August to discuss water-transport matters for the following year. For the rest of the year, the Canal Governor stayed in Huai'an to supervise boats crossing the Huaihe River, while the *Caoyun* Commander stayed in Xuzhou to supervise boats crossing the Yellow River. In the early Qing Dynasty, the Canal Governor lived in Tongzhou, moving to Huai'an later.

Under the Canal Governor, officials in the provinces (called *Duliang Dao* or *Liangchu Dao*) were appointed to assist him in collecting grains. The *Duliang Dao* officials were also responsible for the supervision of boats in the canals and the Huaihe River. During the Ming Dynasty, one *Duliang Dao* official was appointed to each of the 13 provincial capitals. However, during the Qing Dynasty, the *Duliang Dao* officials were only appointed in Shandong, Jiangsu, Anhui, Zhejiang, Hubei, and Hunan; those in the remaining provinces held only part-time positions. A subordinate position, *Jiandui Zhushi*, was set up under the position of *Duliang Dao*. These officials would live at the points from where the grains departed and were responsible for checking and accepting grains. They also travelled with the grain boats to the transit stations in Huai'an and Yangzhou, where the Canal Governor had to inspect the goods.

In the water-transport system, each province represented a unit (*Zong* in the Ming Dynasty and *Bang* in the Qing Dynasty). The water-transport organisation later evolved into one unified gang, namely the *Cao Bang* in the Qing Dynasty and the *Qing Bang* during the Republic of China. According to *Ming Dynasty: Shihuozhi*, the *Caoyun* Commander commanded 120,000 members of the gang in 12 *Zongs*, which is a rather large group of people. During the reign of the Kangxi Emperor in the Qing Dynasty (AD 1662–1722), one *Yunding* and nine sailors were set up for each boat. With a total of 7,692 boats at that time, the number of *Yunding* and sailors reached 76,920. During the reign of the Jiaqing Emperor (AD 1796–1820), the number of boats and crewmembers reduced.

During the Ming Dynasty, the Inspection Censor, a capital official, was the supervisory official for water transportation. The position was held by a eunuch or an Imperial guard. The Censor oversaw two divisions, with authority over the rivers and canals and water transportation. The Censor lived in the transit stations, which were usually in Huai'an during the Ming Dynasty. During the Qing Dynasty, the Inspection Censors were stationed in Huai'an, Jining, Tianjin, and Tongzhou. The garrison in the towns along the canal were responsible for supervising the boats to prevent them from being detained, from carrying smuggled good, and from doing business along the way. After the reign of the Kangxi Emperor, this management was carried out by soldiers along the river according to the flood segments.

1.4.3 *Project Management: Mainly Water Sources and Dams*

During the Ming and Qing dynasties, the Grand Canal was divided into several sections according to its geographical features. During the Ming Dynasty, the canal was divided into seven sections — *Bai Cao*, *Wei Cao*, *Zha Cao*, *He Cao*, *Hu Cao*, *Jiang Cao*, and *Zhe Cao*. After the construction of the Zhonghe River during the reign of the Kangxi Emperor in the Qing Dynasty (AD 1662–1722), the canal crossed the Yellow River in Huai'an before connecting to the Zhonghe River at the north shore of Suqian. *He Cao* no longer existed. Among the six canal sections, *Zha Cao* (the Huitong River) was the most important. The Huitong River crossed the watershed horst in Shandong and the water levels in the canal were mainly regulated by gates and dams. The joints between the canal and the Yellow River, the Huaihe River, and the Yangtze River were the water-transport junctions. These intersections were also affected by changes in terrain and river sedimentation, meaning they were selected as locations for key management projects.

1.4.3.1 *Water-Source Management*

The climate along the Beijing–Hangzhou Grand Canal changed significantly from north to south. Downstream of the Yangtze River, the annual rainfall was approximately 1,200–1,400 mm, approximately 1,000 mm south of the Huaihe River, and approximately 600 mm in the Yellow River and the Haihe River basin. Due to the differences in rainfall across the regions, the constraints of topography, and uneven distribution of water during the wet and dry seasons, the canal was reliant on the regulation of water from the lakes and ponds along the canal to maintain minimum navigation conditions.

During the Ming and Qing dynasties, particular attention was paid to the management of water resources. The collection of spring water and surface water from lakes and ponds formed an important part of the canal's management. In the arid regions, spring water was the main water source for the canal. The facilities were directly managed by the state government. There were special laws for the management; for example, the culprit who broke the embankment of Nanwang Lake in Shandong Province and Zhaoyang Lake in Peixian Town, or the springs of Taishan and other areas, would be sent into exile or to garrison the frontiers or border regions. During the Ming

Dynasty, complex laws were written regarding the management of water sources. People who took water from the springs of Shandong, Nanwang Lake in Jining, Shushan Lake, Anshan Lake, Zhaoyang Lake in Pei County, Gaobao Lake in Yangzhou, or from water sources in Gaojia Dam and Liuwanpu in Huai'an Province without permission would receive heavy penalties. Meanwhile, the main water sources were managed by special officials and labourers, such as *Hu* (lake) *Fu*, *Tang* (pond) *Fu*, *Quan* (spring) *Fu*, and *He* (river) *Fu*, who engaged in patrols and dredging.

In order to meet the needs of water transport, the water sources along the canal were introduced into regulated lakes and ponds, which affected irrigation and domestic water use in the local area. During flood season, water was discharged from the lakes and ponds, which caused waterlogging. The floodplains of the lakes and ponds were good places for agriculture cultivation, which meant a conflict arose between water storage and the usage of land for arable farming. In the Ming and Qing dynasties, the state established the water and land use system for these special water sources and floodplains to ensure that water transport was maintained while also balancing the needs of farmers and citizens.

The water management by the state often violated the interests of local despotic gentry; it was difficult to avoid the encroachment and appropriation of canal water sources. For example, it was a long-term practice to enclose the Lianhu Lake floodplain (south of the Yangtze River) for cultivation. The agricultural fields here were repeatedly ruined to maintain the Lianhu Lake's role as a water tank, but this practice could not be stopped. The lake was totally reclaimed by agriculture during the late Ming Dynasty.

Water had to be introduced into the canal directly through tidal bores. From the reign of the Jiajing Emperor during the Ming Dynasty (AD 1521–1567), a large portion of the Nanwang Lake floodplain along the Huitong River was occupied by the local despotic gentry for cultivation. During the 1st year of the reign of the Yongzheng Emperor in the Qing Dynasty (AD 1723), the areas around the lakes of Nanwang, Ansan, and Zhaoyang were remeasured, and boundaries were established around them to forbid occupation. The lake surface areas gradually increased. During the Ming and Qing dynasties, the strict management of the Huitong River's water sources effectively prevented the encroachment of the local gentry, maintaining the storage tanks and keeping the water running smoothly through the channel for over 500 years.

1.4.3.2 *Sluice and Dam Management on the Canal*

Sluices and dams on the canal could be divided into two types according to their functions — diversion sluices and dams and regulating sluices and dams. The diversion sluices and dams were built on the side of canal and were used to supply water to the canal or drain water from it. The regulating sluices and dams were built across the river, and they were used to control water depth to aid navigation.

The regulating sluices required the most management. During the Tang and Song dynasties, regulating sluices were used at the joint sections of the Jiangnan Canal and the Huaiyang Canal. Due to management difficulties, the sluices were changed to regulating dams in the Ming and Qing dynasties. In the Tonghui, Huitong, and Zhong canals, regulating sluices were still used to control the depth of water along the channel. The key management practices of sluices included the joint use of upper and lower gates (i.e. keeping one open with another closed) and ensuring boats crossed the gates in groups. It was essential that the opening and closing of the gates was timed correctly for the waterway to be utilised efficiently. The amount of time for which the gates were opened and closed depended on the depth of the canal, which was generally maintained at approximately 2 m.

Canal dams were built to reduce the falling heads and maintain a depth suitable for navigation. The difference in water levels between the upstream and downstream was reduced between two dams. On the dam itself, a mechanism involving twisted rope assisted boats in their passage over it. Grains were removed from boats carrying them before workers tugged the boats over the dam.

The river course, canal engineering facilities, and water tanks of the canal were subject to regular and routine maintenance. The regular maintenance included annual repairs and major repairs. River dredging and embankment construction were carried out simultaneously, and this was called *Tiao Jun* work. The annual maintenance was known as 'small *Tiao*' or 'small *Jun*', while the major maintenance was known as 'Big *Tiao*' or 'Big *Jun*'. Major maintenance was carried out either every 2–3 years or 4–5 years according to different sections of the canal. During annual maintenance, water continued to flow through the canal and only the moon river beside the gate was dredged. During construction, boats moved up and down over the main stream of the river. During periods of major maintenance, a dam was built to stop the main stream's flow and the main river channel was dredged. Boats moved through the moon river. The Nanwang section of the Huitong River was the key place for maintenance.

As with the repair of the Yellow River dike, a paramilitary management organisation was implemented by the military garrison for the canal levee repair. Each section of this organisation was given separate roles, such as road paving, regular patrol, and the catching of thieves. During annual maintenance and periods of major maintenance, labourers participated in construction. In the 1st year of the Kangxi Emperor (AD 1662), regulations were put in place regarding canal construction; if the canal levee burst within 3 years of the canal's construction, the officials responsible would be penalised. During the Qing Dynasty, the *Biao Fu* (marking labourers) were hired to manage shoal marks and guide boats up and down the North Canal in the north of Tianjin.

1.4.4 *Water Transport Management: A Paramilitary Transport Organisation*

During the reign of the Tianshun Emperor in the Ming Dynasty (AD 1457–1464), the number of boats available for water transport was 17,770 and the number of officers and soldiers allocated to the transportation of grain was 120,000. From the reign of the Chenghua Emperor to the end of the reign of the Hongzhi Emperor (AD 1465–1505), the number of boats was 11,427 and the number of officers and soldiers was 114,500. During the reign of the Qianlong Emperor in the Qing Dynasty (AD 1736–1795), the number boats reached 17,692; in the 17th year of the reign of the Jiaqing Emperor, the number was 14,092.

During the reign of the Chenghua Emperor in the Ming Dynasty (AD 1465–1487), the required annual amount of grain to be sent to the capital via water transport was 4 million dan.[5] During the Qing Dynasty, the required amount of grain was roughly the same. Each year, the amount of grain collected from Shandong and Henan provinces was approximately 350,000 dan, and it was referred to as 'North Grains'. The annual contribution from Jiangsu was 1.8 million dan, from Huguang (Hubei and Hunan provinces) it was 250,000 dan, and from Zhejiang and Jiangxi it was 1.2 million dan. These were referred to as 'South Grains'.

Among the grains themselves, the Japonica rice and glutinous rice (known as 'white grain') collected from Suzhou, Songzhou, Changzhou, Taicangzhou, Jiaxing, and Huzhou were directly transported to the Imperial Household in Beijing to be used by the Imperial court and nobility.

Besides grains, there was also a large amount of official and commercial material being transported over water. Official transport included boats containing tributes, boats carrying seasonal goods or officials, fast boats carrying post, and boats transporting building materials such as wood, bricks, and minerals. Commercial transportation also accounted for a large proportion of vessels on the canal, over which mostly handmade products and general merchandise were carried.

1.4.4.1 *Water Transport Management: The Combination of Branch River Transportation and Warehouse Storage*

The segment method was implemented for water transport during the Ming Dynasty, including the 'branch method' and the 'exchange method'. This method was gradually eschewed in favour of the direct transport method (also known as the long transport method). During

[5] 1 dan equals 10 dou, which equals 100 sheng. (1 dan = 10 dou = 100 sheng; 1 sheng ≈ 1 litre)

the Qing Dynasty, the direct transport method was inherited with little change.

For the branch method, all the grain was transported to warehouses along the canal and then taken to the capital by the army. Warehouses were built along the canal for grain storage and transport. Major warehouses were located in Huai'an, Xuzhou, Linqing, Dezhou, and Tianjin. Grains from Huguang and Zhejiang were transported through the canal and stored in the Huai'an warehouse. Similarly, grains from Suzhou, Songjiang, and Jiangxi were stored in the Xuzhou warehouse. Grains from Yingtian, Changzhou, Zhenjiang, Huai'an, Yangzhou, Taicang, Chuzhou, and Hezhou were stored in the Linqing warehouse. The grains in the warehouses were transported along the canal to Beijing under military supervision. Through the regulation and organisation of the warehouses along the canals, the branch transport method could adapt to changes in the river water in different sections, and these changes could be made several times per year.

The exchange method represented a shift from civil transport to military transport, and it began in the 7th year of the reign of the Xuande Emperor in the Ming Dynasty (AD 1432). Grains from areas south of the Yangtze River were transported to the warehouses in Guazhou and Huai'an through civil transport before being handed over to garrison units in Fengyang and Yangzhou, north of the Yangtze River. The grains were then transported to Tongzhou or Beijing by the military. Civil transport groups no longer operated over the Huaihe River or the Yellow River. An amount of grain was deducted from the total according to the distance travelled in order to pay for transportation costs. For example, the deducted amount of grain for a trip from Huguang was 8 dou per dan, from Jiangxi and Zhejiang it was 7 dou per dan, and from the south of Jiangsu it was 6 dou per dan. After the implementation of the exchange method, the boats in the south of the Yangtze River stopped going north after crossing the river, eliminating the high cost of going over gates and dams.

In the 7th year of the reign of the Chenghua Emperor (AD 1471), after the reformation of the exchange method, the long transport method was implemented. This meant the grains were transported from beginning to end with no break. Grains from different areas were sent to specified docks and submitted to military officers. Warehouses were built in the south of the Yangtze River to collect 'South Grains', and boats did not need to cross the river. Besides the deductions already taken, an additional amount of grain was taken to account for cost of crossing the river — 6 sheng per dan.[6] Along the canal, except at Jingcang and Tongcang, the rest of the warehouses were used to store grains and provisions for the army. The long transport system continued to be used into the Qing Dynasty.

1.4.4.2 *Grain Transport: Time Management*

In order to ensure that approximately 4 million dan of grains were smoothly transported to Beijing each year, it was essential to control the transport schedules, including the time boats spent sailing and docking, the time sailors spent on land, and the time soldiers spent escorting boats and checking for contraband. The boats used to transport grains to the capital (*Cao Chuan*) had to arrive at their destinations within a specified schedule, and they also had to cross the Huaihe River to arrive at Tongzhou and arrive in Beijing at specific times.

During the reign of the Zhengde Emperor in the Ming Dynasty, a 'water chart' and other relevant documents were provided to each transport boat, and sailors had to fill the chart to record the route they had travelled along each day. Officials inspected the travel schedule at checkpoints. To avoid the obstruction of boats following behind, boats violating the schedules would be stopped in Dezhou and their shipping grains stored in the Dezhou warehouse. Later, the schedule form was organised with *gan zhi* (Chinese characters like 甲, 乙, and so on), which signified months and days. Sailors had to fill the form with their daily mileage and the places they had docked. If they had missed a day of travelling, the reason had to be stated on the form.

---

[6] 1 sheng is equal to approximately 1 litre.

As the management system of the canal gradually improved, so too did the water transport system. During the early Ming Dynasty, the transit transportation method was practiced and Huai'an was the final destination for the provinces south of the Yangtze River. There were no uniform provisions for the water transport system. Boat management began to apply after the implementation of the long transport system.

During the reign of the Yongle Emperor in the Ming Dynasty, most of the grain boats weighed approximately 200 dan and each boat typically carried 100 dan of grains. These boats were compact and flexible and made several round trips per year. After the implementation of the long transport system, most of the grain boats weighed approximately 500 dan, each carrying around 400 dan of grains. The draught limit was approximately 1 m. The boats continued to increase in size, but the total number of boats travelling along the canal reduced until the reign of the Jiaqing Emperor during the Qing Dynasty (AD 1796–1820), when it was only 4,000.

The big boats had to unload their cargo onto lighters at crossing gates. In the early Qing Dynasty, the number of boats was less than 600, but by the end of the reign of the Qianlong Emperor, the number was over 1,500. Besides the official boats, many private boats were also equipped with their own lighters. The increase in the amount of cargo each boat could carry resulted in greater difficulty when crossing gates and dams, slower speeds, and prolonged journeys (see Table 1.1).

1.4.4.3 *Water Transport Management Regulations*

During the reign of the Yongle Emperor in the Ming Dynasty (AD 1403–1424), the entirety of the Beijing–Hangzhou Grand Canal was in use. A large number of management problems had emerged, some of which were difficult to deal with simply by using the water transport system itself. Therefore, legally binding rules and regulations were established that played an important role in maintaining the normal operation of water transportation along the canal.

Table 1.1: The Evolution of Grain Boat Speeds During the Ming and Qing Dynasties

Period	Starting point	Time limit: To cross the Huaihe River	Time limit: To arrive in Tongzhou or Beijing	Notes
The Chenghua Emperor in the Ming Dynasty (1465–1487)	Henan, Shandong, North Zhili		1 May	
	South Zhili		1 July	1 August for boats crossing the Yangtze River
	Zhejiang, Jiangxi, Hubei, Hunan, Guangxi, Guangdong		1 September	
The Jiajing Emperor in the Ming Dynasty (1522–1566)	Provinces to the north of the Yangtze River	December	Early May	1 April from the reign of the Wanli Emperor
	South Zhili	January	July or August	June or July from the reign of the Wanli Emperor
	Zhejiang, Jiangxi, Hubei, Hunan, Guangdong, Guangxi	March	September	August from the reign of the Wanli Emperor
The Shunzhi Emperor in the Qing Dynasty (1644–1661)	Shandong, Henan		1 March	Initially January
	Provinces to the north of the Yangtze River	December	1 April	
	South Zhili	January	1 May	
	Zhejiang, Jiangxi, Hubei, Hunan, Guangdong, Guangxi	February	1 June	

In the 9th year of the reign of the Chenghua Emperor (AD 1473), 17 prohibitions were decreed regarding the management of water transport, tribute transport, official transport, and river courses. These prohibitions included eight rules:

1. The first rule concerned tribute boats carrying seasonal goods (such as fruits and vegetables). The number of tribute boats was 162, each of which carried 30 seasonal items. Other boats were not allowed to be in the same group as the tribute boats, which were allowed to cross the gates as soon as they arrived.
2. The second rule concerned the opening of the gates. For *Caoyun* boats, official boats, and merchant boats, the gate was opened according to the regulated water depths. The boats went through the gate according to the rule of first come, first served. If someone tried to force the gate open, they would be punished. If the manager of the gate extorted money from others or deliberately circumvented the rules concerning the gate, he would be punished.
3. To prevent the abuse of power, officials who were managing rivers, gates, and water transport were supervised by the River Patrol Censor, who would deal with them if they broke the law. Local officials could not interfere in these affairs.
4. The main officials in charge of managing rivers, gates, floods, and dams within the provinces were given standing positions and forbidden from holding other positions at the same time. Specialist labourers were not allowed to be assigned duties beyond their responsibilities; for example, sailors on official and commercial boats could not arbitrarily require the labourers to tow the boats across dams. Similarly, the local authority could not arbitrarily stop providing the materials and funds for canals and rivers or use them for other purposes.
5. It was illegal to take water from the canal's water sources (ponds, springs, rivers, and so on) without permission. Those found guilty of this act would be banished.
6. For each boat, it was necessary to carry 10 dan of local products so that the sailors could exchange them for living materials (such

as wood and salt) along the way. If the sailors overloaded their boats or smuggled goods, the offending property would be confiscated. Sailors doing business along the way and delaying the boat's schedule would be punished. It was forbidden for *Caoyun* boats to carry passengers or privately owned salt; if either was found, they would be removed from the boat and the persons responsible would be punished.

7. The seventh rule concerned passenger restrictions on official boats. For each official boat returning from Beijing, the Ministry of War would provide documents on which the number of the boat, the names of the sailors, and the names of the officials on the boat would be indicated. Passengers found on the boat who were not listed (including officials) would be punished.
8. The eighth rule contained other provisions. For example, the towpaths along the canal could not be encroached upon for the building of houses. It was also forbidden to dispose of bodies in the river. Finally, the use of percussion instruments was forbidden on all boats except for tribute boats.

During the Ming Dynasty, some state laws related to the management of water transport, such as the *Ming Code*, the *Punishment Ordinance*, and the *Zhanfu Ordinance*. During the Qing Dynasty, regulations regarding professional penalties were included in addition to the previous regulations.

During the Ming and Qing dynasties, the bureaucracy and regulations were well-established and strict. However, contradictory policies came from different departments, and there was some chaos in the management. Furthermore, the administrative incompetence of government officialdom made the river and water transport management departments one of the most corrupt of its kind, with bribery running rampant. In *The Draft History of Qing: Shihuo Zhi*, it is recorded that the costs to local government during the transportation process were as high as the value of the grain itself. Although water transport through the canal had brought benefits in terms of the political and economic stability of the country, society also paid a high price.

Chapter 2

The Tonghui Canal

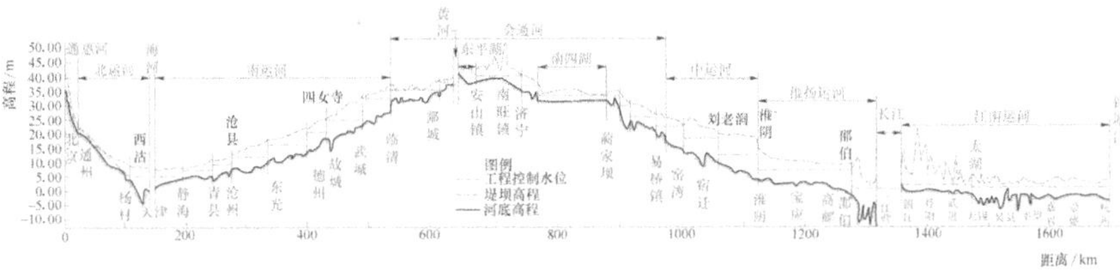

Figure 2.1: Elevation of Different Sections of the Tonghui Canal

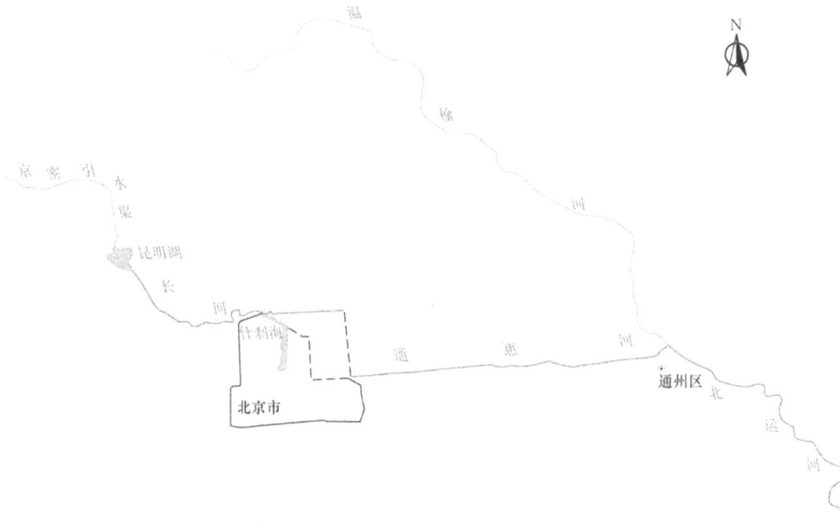

Figure 2.2: The Modern-Day Tonghui Canal

Note: A 21-km-long stretch from Dongbianmen, Dongcheng District, to Beiguan Gate, Tongzhou District.

Due to restrictions caused by terrain, rainfall, and other concerns, problems with the water supply to the Tonghui Canal and the height differences between its two end points caused difficulty to unpowered boats trying to travel along it (see Figure 2.1). To obtain water sources for the canal, control its flow, and improve the transportation along it, comprehensive water-source works and gate assemblies along the route were completed in the Yuan Dynasty. These consisted of a system of canal-water supply and water control. The water-source works on the Tonghui Canal have simultaneously generated a long-term affection for the regional ecological environment and a unique urban style (see Figure 2.2). It has now become part of the Beijing urban river system.

2.1 Basic Information

2.1.1 *Natural Conditions*

The Tonghui Canal is one of the sections of the Grand Canal where the water supply is totally dependent on hydraulic engineering. Beijing is located on the northwest edge of the North China Plain, adjacent to the Western Hills (a branch of the Taihang Mountains) in the west, the Jundu Mountains (the west part of the Yan Mountains) in the north, and the North China Plains to the southeast. Therefore, the overall terrain of Beijing is high in the northwest and low in the southeast. The ground slope of the Beijing urban area is 1–1.2 per cent. The elevations of the urban area in the west and east are generally approximately 50 and 30 m, respectively, above sea level. The network of rivers across the North China Plain are intensive and includes five major river systems — the Yongding River, the Chaobai River, the North Canal, the Juma River, and the Juhe River. Beijing is located at the ridge of a proluvial–alluvial fan of the Yongding River and the Chaobai River. The rivers flow out around the suburbs. There were many springs in the west and north of Beijing, which were the main water sources for ancient canals.

The climate of Beijing is that of a semi-humid and semi-arid monsoon area in a warm temperate zone. The average annual precipitation is 595 mm. The inter-year rainfall and annual distributions are extremely consistent, with the amount of rainfall from June to September taking up around 85 per cent of the annual total. Five major river systems belonging to the northern system of the Haihe River basin are distributed from the east to the west of Beijing — the Ji Canal, the Chaobai River, the North Canal, the Yongding River, and the Daqing River. The runoff of each river is mostly generated during the flood period.

2.1.2 *Historical Development*

In the 9th year of the Zhiyuan Era during the Yuan Dynasty (AD 1272), Kublai Khan established the capital of Khanbaliq in Beijing, and since then Beijing was the political and cultural centre of the Yuan, Ming, and Qing dynasties. During the early Yuan Dynasty, the canal transportation system of the older dynasties in the Central Plains was inherited and continued, and new hydraulic projects were built on a large scale. Firstly, the Jizhou Canal and the Huitong River were excavated to connect the watercourse between the Yellow River and the Yuhe River. The Tonghui Canal was then constructed to enable the direct transportation of tribute grains from the south of Yangtze River to Beijing. Therefore, the basic layout of Grand Canal connecting the south and the north of eastern China was formed.

Before the excavation of the Tonghui Canal, the tribute grains going from Tongzhou to Beijing were transported through the Bahe River. The Bahe River drew its water from the springs of Yuquan Mountain and the Gaoliang River. One section of the upper reach of Tongzhou, through the Baihe River and the Wenyu River, was a natural river course in which the water was frequently too shallow for transportation. In the 28th year of the Zhiyuan Era (AD 1291), after surveying the area and making a comprehensive plan, Guo Shoujing, a scientist and water-conservancy expert, presented a proposal to excavate the Tonghui Canal and begin related water-source works. Kublai Kahn approved the proposal immediately. The project began the next spring and the Tonghui Canal was completed in August of the 30th year of the Zhiyuan Era (AD 1293). From then on, the Beijing–Hangzhou Grand Canal ran from the south to the north (see Figure 2.3).

During the Yuan Dynasty, as the Huitong River was impossible to navigate for most of the year, water transportation mainly depended

Figure 2.3: Route Map of Tonghui Canal during the Yuan Dynasty

Note: During the Yuan Dynasty, the water source of the Tonghui Canal was diverted from Baifu Fountain (located in modern-day Changping District, Beijing). It was merged into other fountains in the foothills of the Western Hills and then flowed into Wengshan Lake and towards the southeast. The water diverted from the Changhe River entered other lakes in the inner city, meaning they could play a role in regulating and storing water. At that time, the Tonghui Canal began in Jishuitan (a remaining part of modern-day Shichahai) and was divided to flow eastward from Haizi Gate (initially known as 'Chengqing Gate' during the Ming Dynasty), reaching Tongxian County through Beijing Donganmen, Zhengyi Road, Taijichang, and directly southeast to the Datong Bridge of the East Side Gate. The water then entered the Baihe River (the modern-day North Canal).

on the ocean. By the end of the Yuan Dynasty, ocean transportation to the south of the Yangtze River was stopped, so the Tonghui Canal was used intermittently. During the early Ming Dynasty, the Tonghui Canal, which had not been maintained for many years, was almost silted up.

During the early Ming Dynasty, the Great Capital of the Yuan Dynasty was renamed 'Beiping'; in the 4th year of the reign of the Yongle Emperor (AD 1406), Zhu Di of the Ming Dynasty renamed Beiping as 'Beijing'. In the 19th year of the reign of the Yongle Emperor (AD 1421), the capital was officially moved from Nanjing to Beijing. From then on, the dredging of the Grand Canal and the reconstruction of the river sections rapidly became urgent concerns.

The city of Beijing was reconstructed in the 7th year of the reign of the Xuande Emperor (AD 1432). The capital's southern wall was moved 2.5 km southwards to what is now Chongwenmen and Zhengyangmen. One northern section of the Tonghui Canal entered the Imperial city; the *cao* ships (which carried tribute rice along the canal) were not allowed to go into the inner city, so the landing dock was changed to Datong Bridge (outside modern-day Dongbianmen), and therefore the Tonghui Canal was known as the 'Datong River' during the Ming Dynasty. The Ming government ordered the excavation of a branch river on the north bank of the Datong Bridge; as such, after reaching Dongbianmen, the *cao* ships would be able to continue north until they reached Dongzhimen and Chaoyangmen.

In August of the 11th year of the Chenghua Emperor (AD 1475), Pingjiang County's Chen Rui took charge of overhauling the Tonghui Canal. The watercourse was altered to be 10 zhang (approximately 30 m) wide throughout the route. In October of the 2nd year, the dredging work of the Tonghui Canal and the gate bases thereof were completed across approximately 30 km of the canal, from Datong Bridge to Zhangjiawan. As the Ming Dynasty's mausoleum was located in Changping during this period, the fountains (such as Baifu) were not diverted during the overhaul. The Xihujing water source in the Western Hills was partially dredged. It was recorded that 'only half [of the] water of the West Lake [was] divided', resulting in a remarkable shortage of water for the Tonghui Canal.[1] Despite the overhaul of the canal, transportation along it had to be stopped after only 2 years, and over the next 40 years almost no *cao* ships arrived in Beijing.

During the 7th year of the reign of the Jiajing Emperor (AD 1528), Censor Wu Zhong took charge of reconstructing all the gates of the Tonghui Canal and Tongzhou stone dams and moving the port of the Tonghui Canal to North Canal southwards from Zhangjiawan. In this

[1] From a Water Conservancy Assembly document — the river-control project fascicule *Annals of the Ming Dynasty*, compiled by Yang Zhengtai, 1983: 164.

way, some of the tribute grains could be unshipped from the Zhangjiawan stone dam, transported a certain distance by road, and then shipped through the Tonghui Canal. This was a sectional delivery method. The rest of the grains were unshipped from an earth dam here and transported by road to each tribute-grain barn in the capital, representing a 'simultaneous transportation by water and by road'. After the successful dredging under the leadership of Wu Zhong, the passage along the Tonghui Canal remained smooth until the late Ming Dynasty.

In AD 1644, the Qing government toppled the Ming government and the Shunzhi Emperor began his reign. Beijing was chosen as the capital. The canal transportation system established in the Ming Dynasty was essentially continued; the Emperor attached great importance to it, and the canal became an important channel through which the grains and the goods of southeast China were transported to Beijing. The Emperor also undertook a tour of the south. During the early Qing Dynasty, canal transportation was further optimised by the repair of the stone dam and the five gates of the Tonghui Canal (see Figure 2.4).

Figure 2.4: The Navigation of Tonghui Canal and the Application of Ship Locks — *Water Conservancy Map*, Drawn by Hong Wu during the Qing Dynasty

In the 14th year of the reign of the Shunzhi Emperor (AD 1657), *Jishizhong* (a type of official) Lei Yilong presented a proposal to repair the stone dam and the embankments of the five gates before dredging the Tonghui Canal, which had been neglected for a number of years.

In the early years of the reign of the Kangxi Emperor, the watercourse of Jade Fountain Hill, the upper water source of the Tonghui Canal, was successfully dredged, and the gate of the Datong Bridge was reconstructed. In the 35th year of the reign of the Kangxi Emperor (AD 1696), the river course of the Huitong Canal and the five gates thereof were successfully dredged, the embankments were reinforced, and rolling dams were constructed for water-discharge. In the 2nd year, the City Moat was cleared, the Chaoyangmen water gateway and Dongzhimen water gateway were added, and the water channels of the East City Moat and the Tonghui Canal were made available. The *cao* ships were thereby able to enter the East City Moat through Datong Bridge to directly reach Chaoyangmen and Dongzhimen and go ashore near the East Tribute Barn in the capital; therefore, the route across land from the unshipping place to Datong Bridge was shortened. This navigable section of the city moat was roughly parallel with a northern branch of the Yuehe River of the Tonghui Canal.

The tribute grains carried by the *cao* ships going north were unloaded in Tongzhou to be stored in a Tongzhou barn or transferred to Beijing to be stored in a barn there. The tribute grains transferred to Beijing were shipped by local vessels along the Tonghui Canal in the north stone dam of Tongzhou. They were then transferred gate-by-gate until they reached Datong Bridge, at which point they went north and finally arrived at the Beijing barn through the East City Moat. Due to a limit on the amount of grains that could be transported along the canal, the road from Tongzhou to the capital had to be built for land transportation.

During the Qing Dynasty, the water source of the Tonghui Canal was not diverted from the water of Baifu Fountain, but was instead collected from the Yuquan Fountain in the Western Hills. Additionally, during this period, the rolling dam of the Tonghui Canal and the Yuehe River were used the most in pursuit of a good canal storage-discharge relationship.

During the reign of the Xianfeng Emperor (AD 1851–1861), most of the grain-production areas south of the Yangtze River were occupied by the Taiping Heavenly Kingdom. Because of this, the progress of the Beijing–Hangzhou Grand Canal was interrupted for almost 10 years. The canal soon recovered, but the situation gradually deteriorated. In the 26th year of the reign of the Guangxu Emperor (AD 1900), the Eight-Nation Alliance invaded Beijing and took all the grains from each of the capital's barns. The following year, the Beijing–Tianjin was opened to traffic; with the advantages of railway transportation becoming increasingly clear, transportation along the Beijing section of the canal came to a forced stop. The Beijing–Tianjin ship route had been planned in the 1820s, but it was not implemented. The gate dams of the Tonghui Canal were gradually abandoned and the river channel was silted up so as to evolve into the Beijing drainage channel.

2.2 Critical Work

2.2.1 *Water-Source Work*

Each river system in the Beijing area had the typical characteristics of northern rivers — severely inconsistent annual distribution of runoff water, high sand content, greater river slope, and the necessity of water-source work.

The water-source work planned by Guo Shoujing (see Figure 2.5) on the Tonghui Canal was formed of three parts: fountain-diversion work from Baifu Fountain to Wengshan Lake (the modern-day Summer Palace and Kunming Lake); water-storage work on Wengshan Lake and other lakes within the city; and water-diversion work on the Changhe River. Water from the remains of Kunming Lake, Zizhuyuan Lake, Zoo Lake, Shichahai, Beihai, and Zhongnanhai is still kept in many water tanks even today. The presence of these lakes meant that the adjustment of annual water requirements could be essentially realised, so Guo Shoujing's plan successfully enabled the adequate and stable supply of the Tonghui Canal with water. The water-source work on the Tonghui Canal became the river and lake system of modern-day Beijing, creating lasting benefits for the city's development.

Figure 2.5: The Former Site of the Xishan Baifu Fountain as Implemented by Guo Shoujing (2009)

The Tonghui Canal is chiselled and dredged along tracks left by the people of the Yuan Dynasty; as such, Shenxian Fountain and Mayan Fountain are broken over, the branches of the Yuhe River and the Shahe River are interrupted, and other fountains merge together to form Qili Lake (Wengshan Lake), which runs eastward through the capital. This canal directly reaches Gaolizhuang through the Datong Bridge below and is connected to the Baihe River. The whole 160 li (80 km) long route is provided with 24 gates.

The divided water of Baifu Fountain in the North Mountain flows southward after turning in the west. The water accumulates each year in Jishuitan after entering the city from the West Water Gate through Wengshan Lake, continues to flow southward after turning in the east, and merges into the old grain transportation river after coming out from the South Water Gate. There is one gate every 10 li (5 km), and there are seven gates in total on the way to Tongzhou. Another sluice gate is found several miles away from each gate, and they are controlled by each other to control the passage of ships and the blockage of water.

Northern Changping and the northwest Western Hills of Beijing (located on the ends of the Yan Mountains and Taihang Mountains, respectively) are not only a transition zone between the Shanqian Plain and the North China Plain, but also a zone where groundwater overflows, where many fountains are exposed and merge into the western Gaoliang River and the eastern Wenyu River.

After making a field visit and planning carefully, Guo Shoujing, the Yuan Dynasty water conservancy expert, divided Baifu Fountain and Xishan Fountain (see Figure 2.6) at Changping to form depression lakes within the Summer Palace area through water-diversion

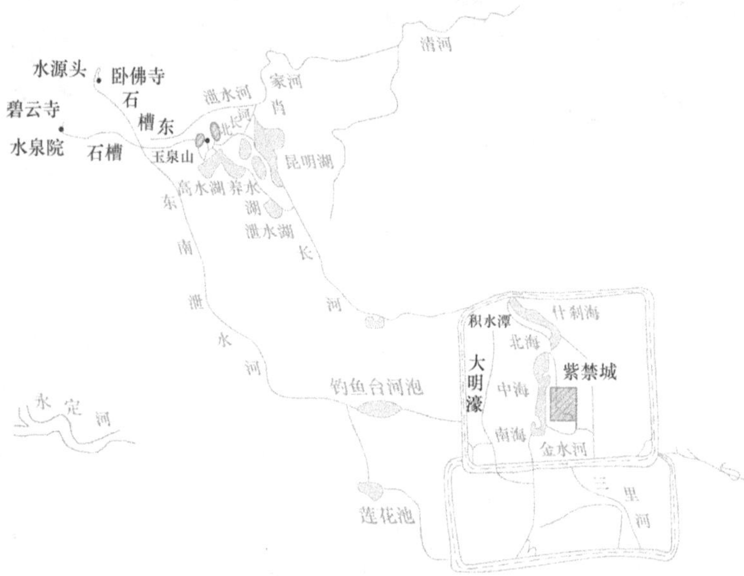

Figure 2.6: A Water-Source Work on the Tonghui Canal in the Qing Dynasty

Note: The Xishan Fountain, including the fountains on the modern-day Jade Fountain Hill, was the upper source of the Gaoliang River. The Xishan Fountain and Jade Fountain were developed as the transportation-supporting water sources for the Bahe River. During the Ming and Qing dynasties, Jade Fountain Hill was still an abundant outlet of fountain water; during the reign of the Qianlong Emperor, the water sources of Xishan Cherry Valley and Biyunsi were mainly diverted into Kunming Lake as the fountain water.[2] The water in the Xishan Fountain gradually reduced from the beginning of the 20th century; in December 1951, the water yield was 1 m^3/s, but the fountain was basically dry by the end of the 1960s.[3]

---

[2] *Current Old News Record: Vol. 101*, compiled by Yu Minzhong *et al.* (Qing Dynasty), 1983: 1672.

[3] *Beijing Ancient Canal and Urban Water Supply Development*, compiled by Cai Pan and released by Beijing Publishing House, 1987: 88.

work. He also constructed dams for water storage to form regulating water tanks. These water tanks were known as 'Wengshan Lake' during the Yuan Dynasty, 'West Lake Landscape' during the Ming Dynasty, 'Kunming Lake' during the Qing Dynasty, and later 'Gaoshui Lake' (the modern-day Tuancheng Lake). On the lower reaches of the water-storage works at the water source, Guo Shoujing renovated the old route of the Gaoliang River into the Long River, a canal diverting water to the Tonghui Canal. There were water tanks along the route of the Long River from Kunming Lake to the Tonghui Canal, mostly including sporadic lakes, water-storage ponds, and the likes of Zizhuyuan and Zoo from northwest to southeast. This was a perfectly planned water-source work, and it became the core of the present-day urban river and lake system in Beijing.

1. Water-Source Work: Fountain Diversion Work

Baifu Fountain was the upper water source for the Sand River in the Wenyu River system. Groundwater overflowing zones and fountains were spread across the village of Baifu in Changping and the surrounding areas. Baifu Fountain was not only the first divided fountain, but also the most famous. There were dozens of other famous fountains that were divided, including Shenshan Fountain, Wangjiashan Fountain, Xihuyan Fountain, Yimu Fountain, and Tanglong Fountain.[4]

The Baifuwengshan River went southwest from Baifu Village and entered Wengshan Lake from the Qinglong Bridge Gate. The Baifuwengshan River was crossed by multiple mountain streams of the Wenyu River system along the route. Eleven level-crossing works were built at the point where each fountain entered the river along the route.[5] These level-crossing works were equivalent to modern-day overflow weirs and were formed by filling and stacking block stones in braided twig cages. The water could be stopped and collected into the canal within a short time, but these works were frequently broken

[4] *The Total Annals of the Yuan Dynasty*, Vol. 1, compiled by Bei Lanxi *et al.* (Yuan Dynasty), released by Zhonghua Book Company and collected by Zhao Wanli, 1966: 15.

[5] 'History of the Yuan Dynasty: River and Canal Annals I' in *The Twenty-Five-Dynasty River and Canal Annals* (ed.): 244.

during the flood season. This water channel was used until the 1950s and is still partially used as the present Jingmi Diversion Canal today.

2. Water-Storage Work: Kunming Lake and the Work Appliance There

Kunming Lake was known as 'Wengshan Lake' in the Yuan Dynasty and 'West Lake' in Ming Dynasty. In the 15th year of the reign of the Qianlong Emperor during the Qing Dynasty (AD 1750), the Qingyi Gardens (soon to be renamed as the 'Summer Palace') were built southeast of West Lake, which was renamed as 'Kunming Lake' (see Figure 2.8) later that year. Kunming Lake, the most important water-storage work on the Tonghui Canal, faced Longevity Hill, which was in the Summer Palace, to the north, was embanked in the south, and was provided with a water-release gate to control the amount of water inside.

In the 29th year of the Yuan Dynasty, Guo Shoujing diverted the water of Baifu Fountain and other fountains across dozens of miles in the northwest of Jade Fountain Hill into the lake, so the lake's surface basin was expanded due to this great water yield. In the early Ming Dynasty, the Baifuwengshan River was cut off and the water within the West Lake was left to entirely depend on the collected fountain water from the Jade Fountain Hill.

A large-scale expansion project was begun during the reign of the Qianlong Emperor, mainly consisting of the addition of water sources and the enhancement of regulation capability. Consequently, a Kunming Lake key water-control project was formed. In the 14th year of the reign of the Qianlong Emperor (AD 1749), the lake was expanded eastwards and southwards so as to reach the east of modern-day Zhichun Bridge in the east and modern-day Xiuyi Bridge from South Lake Island in the south. The central island in the middle of the lake was connected with the east bank by a 17-hole bridge. In the east, a west embankment was built by imitating the Hangzhou West Lake, and six bridges were built on it so that the lake was connected from east to west. In the 16th year of the reign of the Qianlong Emperor (AD 1751), a total of 7 km of water-division grooves were laid and the Xishan fountain water (see Figure 2.7) of Xiangshan,

Figure 2.7: The Ruin of the Xishan Water Division Stone Groove in Qing Dynasty (1980s)

Note: Exposed underground areas of fountain water and mountain flood were collected into the Cherry Valley and square ponds of temples and then diverted into Kunming Lake through the Water Division stone groove. Some residual sections of the Xishan stone groove exist even today.

Wofosi, and so on were merged therein. Furthermore, with the construction of the west embankment by imitating the Hangzhou West Lake and its six bridges, the lake was cut off from east to west to form gradient water-storage form.

During the Qing Dynasty, Gaoshui Lake and Yangshui Lake (the modern-day Tuancheng Lake and Fengchan Lake) adjacent to West Lake were successively constructed according to their terrain. Together, they formed a work system for regulating water storage together with Kunming Lake. Today, Tuancheng Lake, a terminal balance tank into which south-to-north divided water flows after reaching Beijing, represents a hub for the combined dispatching of water from the Beijing Miyun Reservoir and south-to-north divided water.

The highest elevation of the embankment was approximately 51 m, generally 1–1.5 m higher than the water level. The embankment from Wenchangge to Xiuqiao Bridge was approximately 1,800 m long and approximately 22 m wide on average. The water-facing surface was constructed and protected with boulder strips, and the back was filled with cement. The east wall of the modern-day Summer Palace is situated on the embankment, and the embankment structure can be seen clearly outside the wall.

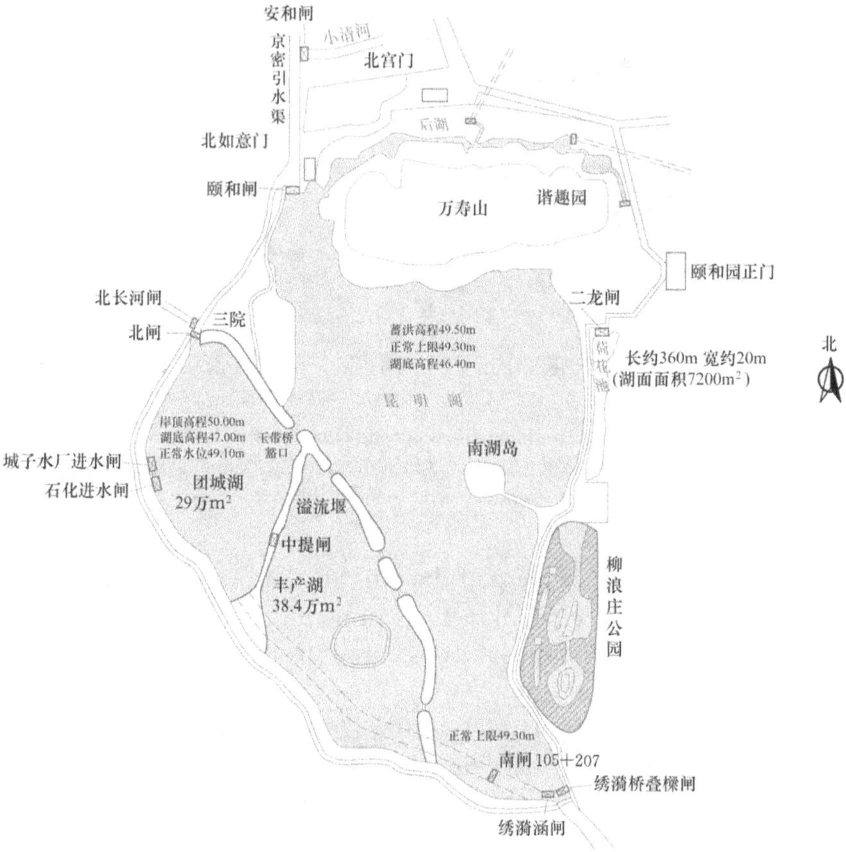

Figure 2.8: A Schematic Diagram of the Main Projects on Kunming Lake

Note: Kunming Lake had a similar system to modern reservoirs, consisting of embankments, a water-inlet gate, a check gate, and a water-supply gate. It was a large, key water-control project that was used for irrigation, storage, and discharge. The Kunming Lake and its main works (2010) basically retained the basic forms and work appliances they had had since the Qing Dynasty.

The west embankment was built to block and store the water from Jade Fountain Hill. Kunming Lake (see Figure 2.9) was divided into two major parts by the west embankment; the part to the east of the embankment was called 'Kunming Lake' (see Figure 2.10), but the other part, to the west of the embankment, was still called 'West Lake'. Six bridges were built on the west embankment to connect the

Figure 2.9: Jade Fountain Hill — Kunming Lake in 1927

Note: The constructed west embankment mainly aimed at storing water in divided areas and the landscape of Hangzhou west embankment was copied. The northwest part of the lake was silted due to insufficient design elevation of stored water so as to be gradually developed into rice field.

Figure 2.10: Kunming Lake Basin and East Embankment (2009)

Note: Kunming Lake had a circumference of approximately 15 km and an area of approximately 2.2 square km in 2010.

two lakes. It was estimated that 2.6 million cubic metres of water could be stored according to the 130-hectare area of the lake during the beginning years of the Republic of China, with an average depth of 2 m.[6]

3. The Qinglong Bridge Gate

A water-discharge gate on Kunming Lake was located close to the north wall of the Summer Palace (see Figure 2.11).

Figure 2.11: The Qinglong Bridge Gate in the 1890s

Note: When the lake's water level was higher than 50 m, a three-hole gate under the Qinglong Bridge could be opened, causing the lake water to be rapidly discharged into the Xiaojiahe River and the Qinghe River. This would ensure the integrity of the east embankment, meaning the urban area and nearby farmlands were protected from flooding.

4. The Erlong Gate

The Erlong Gate (see Figure 2.12) was described thus: 'the east embankment of Kunming Lake was provided with a gate and culverts, which can be opened or closed at any time'.[7]

[6] *Beijing Ancient Canal and Urban Water Supply Development*, compiled by Cai Pan and released by the Beijing Publishing House, 1987: 83.

[7] *Current Old News Record: Vol. 84*, compiled by Yu Minzhong *et al.* (Qing Dynasty), 1405.

Figure 2.12: Erlong Gate (2003)

Note: The Erlong Gate was a two-hole gate, with each hole being approximately 2. 5 m wide. The flashboard was generally not lifted, and the downstream water entered the Old Summer Palace across Machang Bridge after flowing through Yingshi Street. Four culverts were located in the south of the Erlong Gate, used for the irrigation of rice fields in the Liulangzhuang zone before passing across Hongqiao Bridge, through west Weixiuyuan, and entering the Qinghe River by bypassing Qinghuayuan.

5. The Xiuyi Gate

The Xiuyi Gate (see Figure 2.13) was located on the south end of Kunming Lake as a water outlet for the lake and a water inlet for the Long River, and it controlled the distribution of the water to the Long River. Besides the Xiuyi Gate, Kunming Lake also had many other water outlets. An investigation showed that the Houxi River had three outlets at the Xiequ Gardens.[8] The first outlet was located near Tiaoyuan House; after coming out through the palace wall, the water through the first outlet flowed to Chuihong Bridge and reached the Old Summer Palace. The second outlet was located at Xiequ Gardens and was called the 'Northeast Huishan Garden Gate'. After coming out through the palace wall, the water went through the second outlet and went east, being collected into the Summer Palace in front. The third outlet was called the 'West Gate of the Xiequ Gardens'; after coming out through the palace wall, the water went through the third outlet and went south, merging with the water that came out from the Erlong Gate (see Figure 2.14)

[8] *Beijing Ancient Canal and Urban Water Supply Development*, compiled by Cai Pan and released by Beijing Publishing House, 1987: 84–85.

Figure 2.13: The Xiuyi Gate and a Bridge (2010) at the Port of Kunming Lake

Note: This is the Xiuyi Gate and a bridge at the port of Kunming Lake, wherein the water outlet of Kunming Lake and the water inlet of the Long River controlled divided water to Long River.

Figure 2.14: A Gate Excavated Approximately 400 m Outside the South Gate of the Summer Palace

Note: At the end of May 2003, during the widening of a new palace gate road at the south gate of the Summer Palace, approximately 400 m from the south gate itself, a sluice was discovered that had been buried underground for many years. It was 2.54 m wide and one wood plate of the gate had been left inside. It was judged that the sluice had been built during the Qing Dynasty, possibly according to a style similar to that of the Erlong Gate in the Summer Palace (the sluice was refilled on 8 June). This picture shows a view of the sluice from the west.

after passing through the East Palace Gate (two white jade bridges existed in the south and north). An investigation carried out in the 17th year of the Republic of China (AD 1928) showed that Kunming Lake kept six water outlets until the 1920s.

6. Water-Division Work: The Long River

The Long River, which was formed during the process of dredging the old route of the original Gaoliang River, was known as the 'Zaohe River' during the Jin Dynasty, the 'Jinshui River' during the Yuan Dynasty, the 'Jade River' during the Ming Dynasty, and the 'Long River' since the reign of the Qianlong Emperor during the Qing Dynasty. The Long River had a total length of 10.8 km from the Xiuyi Gate of Kunming Lake to the outside of the Xizhi Gate across Changchun Bridge, White Stone Bridge, and Gaoliang Bridge. Along the route of the Long River, the depressions of the former Gaoliang River were developed into water tanks, and the lakes of the modern-day Purple Bamboo Park (see Figure 2.15) and Zoo are the remaining basin of the Long River's water tanks. The ruins of the Maizhong (zhuang) Bridge, the Guangyuan Gate, the Gaoliang Gate

Figure 2.15: One Section of the Long River within Purple Bamboo Park in 2013

Figure 2.16: The Qing Docks in Kunming Lake (2006)

Note: The Qing docks in Kunming Lake were built during the reign of the Qianlong Emperor and are now vital evidence of the navigation of the Long River in those years.

(see Figure 2.19) and so on were still maintained on the Long River. In the 14th year of the reign of the Qianlong Emperor, after the overhaul of Kunming Lake and the Summer Palace, the latter became a main royal pavilion, the Long River (see Figure 2.16) became a main waterway stretching from the Forbidden City to the Summer Palace and Kunming Lake, and temples and wharfs were successively built along the river.

Not only did the Long River play a part in supplying water to the Tonghui Canal, but it was also a main source of water for the royal gardens in the capital. For this reason, the course of the river and its water source were managed strictly. The Long River was provided with watermills during the Ming Dynasty, and during the Qing Dynasty it was forbidden for civilians to take water from it.

2.2.2 *Water Course and Control Work*

The longitudinal slope of the ground between Beijing and Tongzhou was approximately 1 per cent. In the presence of continuous check gates along the Tonghui Canal, the obstruction caused by the difference in terrain elevations was solved for the heavily laden *cao* ships. On the other hand, the water current was also effectively controlled and the limited water sources were made to stretch as far as possible.

When planning the gate bases along the Tonghui Canal, Guo Shoujing originally foresaw the construction of 20. Due to the greater topographical gradient, the Chengqing Mid-Gate (see Figure 2.20) and Pingjin Mid-Gate were soon added. The two gates were shaped in a combined form of three upper, middle, and lower gates, plus the Guangyuan upper and lower gates (see Figure 2.18) that had been built 3 years previously. As a result, the Long River and the Tonghui Canal were provided with 24 gates at 11 locations during the Yuan Dynasty (see Table 2.1). As the water sources gradually decreased in

Table 2.1: Locations of the 24 Gates on the Tonghui Canal in the Yuan Dynasty

Gate		Location	Space	Remark
Guangyuan Gate	Upper gate	In the west of modern-day Zizhuyuan, 70 m from the east of Wanshousi	1,000 m	
	Lower gate	Near White Stone Bridge to the west of the modern-day Capital Gymnasium		
Xicheng Gate	Upper gate	On the west side under Gaoliang Bridge, outside Xizhimen		Renamed 'Huichuan Gate' in the 1st year of Yuanzhen governance (AD 1295)
	Lower gate	On the east bank of the City Moat and on Gaoliang River on the west side of the North Water Gate		

(*Continued*)

Table 2.1: (*Continued*)

Gate		Location	Space	Remark
Chaozong Gate	Upper gate	In the south of Wanyiku and the east of Yimen and the North Water Gate	~200 m	
	Lower gate	To the west of Haizi		
Haizi Gate	Upper gate	On the west side of Backdoor Bridge	~540 m and 1 m height difference	Renamed 'Huichuan Chengqing Gate' in the 1st year of Yuanzhen governance (AD 1295)
	Middle gate	At the south port of the modern-day East Buyaqiao Lane	~500 m and 1m height difference	
	Lower gate	On the east end of the modern-day Beihe Lane		
Wenming Gate	Upper gate	Southeast of the northern crossing of the modern-day Zhengyi Road		
	Lower gate	Between two lanes of Taijichang		
Weicun Gate	Upper gate	In the east of modern-day Chuanban Lane		Renamed 'Huihe Gate' in the 1st year of Yuanzhen governance (AD 1295)
	Lower gate	Southeast of the modern-day Beijing Station		
Jidong Gate	Upper gate	The site of the Qingfeng Gate (Figures 2.17, 2.21) outside the modern-day East Side Door	~1, 800 m	Renamed 'Qingfeng Gate' in the 1st year of Yuanzhen governance (AD 1295)
	Lower gate	Near Shengou Village		

(*Continued*)

Table 2.1: (*Continued*)

Gate		Location	Space	Remark
Jiaoting Gate	Upper gate	The modern-day Gaobeidian Gate	~2,520 m	Renamed 'Pingjin Gate' (Figure 2.22) in the 1st year of Yuanzhen governance (AD 1295)
	Middle gate	No reference		
	Lower gate	Modern-day Garden Gate Village		
Yangyi Gate	Upper gate	The Tongzhou West Gate (modern-day Puji Gate Village), of which the ruins are still extant	Approx. 1,800 m	Renamed 'Puji Gate' in the 1st year of Yuanzhen governance (AD 1295)
	Lower gate	In the east of Laolongbei Village		
Tongzhou Gate	Upper gate	Near the intersection between modern-day Tongzhou Xinhua Avenue and Renmin Road		Renamed 'Tongliu Gate' in the 1st year of Yuanzhen governance (AD 1295)
	Lower gate	Outside the Tongzhou South Gate and in a place that was known as 'Nanpu Gate' in the Ming Dynasty		
Hemen Gate	Upper gate	2 miles from the north of Zhangjiawan City and near Tuqiao Bridge		Renamed 'Guangli Gate' in the 1st year of Yuanzhen governance (AD 1295)
	Lower gate	To the east of southeast Hegezhuang, Zhangjiawan		

the late Ming Dynasty, only five gates along the Tonghui Canal were in use from the 16th century to the 19th century; furthermore, the gates were not opened, thereby only playing a role in controlling water and storing water.

Figure 2.17: Qingfeng Gate in the Ming and Qing Dynasties[9]

Figure 2.18: The Site of Guangyuan Upper Gate (2010)

[9] *Clear Cause Map on Two-Gate Repair*, Vol. I, compiled by Lin Qing, Wang Chunquan, *et al.* (Qing Dynasty) and released by the Beijing Chinese Classics Publishing House, 1984.

Figure 2.19: The Site of Gaoliang Gate (2010)

Figure 2.20: The Site of Chengqing Middle Gate (2010)

Figure 2.21: The Site of Qingfeng Gate (2010)

Figure 2.22: The Site of Pingjin Upper Gate (2010)

2.2.3 *Junction*

The Tonghui Canal stretched east to Tongzhou and was connected to the North Canal so that the tribute grains could directly reach the capital. As the North Canal was formed by the dredging of natural rivers, it was greatly affected by changes in the rivers' courses. The connection with the artificially excavated Tonghui Canal was critical for smooth transportation along it. The location of the river mouth of the Tonghui Canal was also changed, particularly in the North Canal (the Baihe River; see Figure 2.23).

Figure 2.23: The Modern-Day Tonghui Canal Merging with the North Canal

During the Yuan Dynasty, the Tonghui Canal, reaching the neighbour of the Tongzhou West Sluice along the old route of the Jinzha River, was changed to pass southeast through the city centre. It then flowed southeast through Liersi, Gaolizhuang, and finally merged into the Baihe River.

As the Yuan Dynasty ended and the Ming Dynasty began, the course of the Baihe River was changed to go north from Zhangjiawan, pass through Shangmatou Village, and reach Tongzhou. It was a long, tortuous path, but this situation did not change until the Jiaqing Era.

In the 16th year of the reign of the Yongle Emperor during the Ming Dynasty, granaries were initially constructed in Zhangjiawan, and from then on, upper, middle, and lower wharfs were gradually constructed. All the ships going upstream to Tongzhou through the North Canal had to pass through here. As such, the function of the Tonghui Canal's watercourse in the south of Tongzhou City was gradually weakened; in the 7th year of the reign of the Jiajing Emperor (AD 1528), the Tongzhou stone dam was built in the north of the city, thereby completely abandoning the watercourse in the south of the city.

In the 36th year of the reign of the Kangxi Emperor during the Qing Dynasty (AD 1697), the rolling dam was built on the north bank of Hulutou; by the 40th year of the Qianlong Era (AD 1775), Hulutou was provided with two rolling dams to guarantee no threat of floods to the stone dam. In this way, the north branch of the watercourse finally became the trunk river course of the Huitong Canal. The urban water course became so unnecessary to navigate that it gradually disappeared.

During the later years of the reign of the Jiaqing Emperor, the main river course of the Tonghui Canal still passed through the Shandong water gateway (an urban through-flow gate) and could reach the earth dam with the dredging of one section of the north river course. All the exchanged grains were shipped by carriage from the earth dam into the middle and west barns of Tongzhou or shipped by carriage to the Chaoyaomen barn along a stone road leading to the capital. Three grain-carrying lanes opened in the 7th year of the reign of the Jiajing Emperor were also used to transfer grain from the *cao* ships into barns for storage.

In the 22nd year of the reign of the Wanli Emperor in the Ming Dynasty (AD 1594), Yu Shiqian, an official of the Ministry of Revenue in Feudal China, drew the water of the Tonghui Canal so it flowed south from the outside of a new city gate in the west and passed through a new city gate in the south; the water at the old city gate in the south and that of the east water gateway merged. This time, the river was dredged for approximately 3,300 zhang (5,495 km) and deepened by 2 chi (0.66 m, chi is a unit of length). This meant that the depth of the river had been extended by an average of four times its previous value. The result was the Tongzhou South City Moat. Now, the *cao* ships were able to carry the tribute grains from the earth dam and over this water channel to each barn in Tongzhou. At one time, the Qing government assigned eight local ships for this transportation. To prevent flooding, one rolling dam was built to the east of the old South Gate, and this was the third route through which the Tonghui Canal was connected with the North Canal. During the early Qing Dynasty, the Tuqiao and Guangli gates on the south branch of the watercourse were neglected so badly that the watercourse became unnavigable; it gradually became silted up, like the modern-day Yudai River.

2.2.4 *The Arrangement and Management of the Beijing Tribute-Grain Barns*

Since the Jin Dynasty, the tribute grain barns were initially arranged in Tongzhou and Zhongdu to store the grains to be carried to the capital. The volume of canal transportation during the Yuan Dynasty was many times that of the Jin Dynasty, and the tribute grain barns were obviously increased in quantity and scale. After the completion of the Tonghui Canal in the 28th year of the Yuan Dynasty (AD 1291), a total of 22 capital barns — including a total of 1,303 chambers — were successively built; 3.29 million dan of grains could be stored, and the storage capacity approached the total annual amount of tribute grains. The Tongzhou Barn was built in the 3rd year of the Taizong Era (AD 1231), initially located in the north of the city. By the end of the Yuan Dynasty, around 50 tribute grain barns had been completed in the capital city, falling under the management of the canal transportation department there.

During the reign of the Yongle Emperor during the Ming Dynasty, after the capital was established in Beijing, 37 barns were completed in the city. In the 2nd year of the reign of the Zhengtong Emperor (AD 1437), after an exchange-transportation method was implemented, the amount of grains that needed to be stored was reduced.[10] However, Beijing and Tongzhou had more grains to store; therefore, one third of the tribute grain barns in Linqing, Dezhou, and Hexi were reduced, and the barns in Beijing and Tongzhou were increased. In the 4th year of the reign of the Jiajing Emperor (AD 1525), a barn office was established to manage the tribute grain barns in Beijing (see Figures 2.24, 2.25). At that time, there was a

Figure 2.24: The Existing Beijing Nanxin Barn (2008)

Note: As an official royal barn for the storage of Imperial grains and pension rice in the capital of the Ming and Qing dynasties, this was located at No. 22 East 40th Lane, Beijing. Construction of the barn began in the 7th year of the reign of the Yongle Emperor during the Ming Dynasty (AD 1409), and its design was based on the Beitai barns of the Yuan Dynasty; it has lasted for 600 years.

[10] The exchange transportation rule: The tribute grains from various places are transported to Huaian and Guazhou and then exchanged with the transportation arm for transfer. In Henan Province, for example, the tribute grains are exchanged with the general ocean transportation agency of Zheyang, Damingfu; in Shandong, the tribute grains are exchanged with the army for transfer in Jining. The expenses involved in transportation incurred by the arm are the responsibility of the common farmers.

Figure 2.25: The Outer Wall of the Beijing Nanxin Barn (2008)

total of eight barns in Beijing. Meanwhile, the tribute grain barns in Tongzhou stored approximately 3.2 million dan of grains, and the barns in the capital stored approximately 2 million dan of grains. The barns in the capital stored fewer grains than those in Tongzhou, contrary to the situation during the Yuan Dynasty.

The Qing Dynasty continued to use the tribute grain barns (see Table 2.2) from the previous dynasty. The Kangxi and Yongzheng Emperors continued to build barns, so Beijing and Tongzhou reached a total of 17.

Table 2.2: The Construction of all Barns in Beijing and Tongzhou during the Qing Dynasty

Barn	Chamber quantity	Construction began	Stored grain	Present address
		Seven Barns in the Capital		
Haiyun	100	In the 26th year of the reign of the Hongwu Emperor in the Ming Dynasty (1393)	Black soya bean and wheat	South Small Street, Dongzhimen

(*Continued*)

Table 2.2: (*Continued*)

Barn	Chamber quantity	Construction began	Stored grain	Present address
Lumi	57	In the 7th year of the reign of the Yongle Emperor in the Ming Dynasty (1407)	Wheat	South Small Street, Chaoyangmen
Nanxin	76	In the 7th year of the reign of the Yongle Emperor in the Ming Dynasty (1407)	Wheat	North Small Street, Chaoyangmen
Beixin	85	In the 7th year of the reign of the Yongle Emperor in the Ming Dynasty (1407)	Wheat	South Small Street, Dongzhimen
Taicang	83	In the 7th year of the reign of the Xuande Emperor in the Ming Dynasty (1432)	Black soya bean and wheat	North Small Street, Chaoyangmen
Fuxin	64	The Ming Dynasty	Wheat	Sipailou, Dongzhimen
Xingping	81	The Ming Dynasty	Wheat	Sipailou, Dongzhimen
		Five Barns Outside the Capital		
Taiping	86	In the 1st year of the reign of the Yongzheng Emperor in the Qing Dynasty (1723)	Wheat	South City Root, outside Chaoyangmen (each barn was provided with a door)
Wan'an East Barn	40	In the 4th year of the reign of the Qianlong Emperor in the Qing Dynasty (1739)	Black soya bean and wheat	North City Root, outside Chaoyangmen
Wan'an West Barn	45	In the 4th year of the reign of the Qianlong Emperor in the Qing Dynasty (1739)	Black soya bean and wheat	East of Shitouzui, outside East Side Door

(*Continued*)

Table 2.2: (*Continued*)

Barn	Chamber quantity	Construction began	Stored grain	Present address
Yufeng	63	In the 6th year of the reign of the Qianlong Emperor in the Qing Dynasty (1739)	Wheat	South of East Road, Reception Office, outside East Side Door
Chuji	108	In the 6th year of the reign of the Qianlong Emperor in the Qing Dynasty (1739)	Black soya bean and wheat	East of North Stone Road, Yufeng Barn, outside East Side Door
		Two Barns in Qinghe and Anhe		
Benyu	30	In the 46th year of the reign of the Kangxi Emperor in the Qing Dynasty (1707)		Qinghe Town
Fengyi	30	In the 7th year of the reign of the Yongzheng Emperor in the Qing Dynasty (1729)		Anhe
		Two Barns in Tongzhou		
West Bar	98	During the reign of the Yongle Emperor in the Ming Dynasty (1403–1424)	Fine rice	East of Tongzhou Tianhuapan Building
Middle Bar	124	During the reign of the Yongle Emperor in the Ming Dynasty (1403–1424)	Fine rice	South of Tongzhou Tianhuapan Building
South Bar	81	Initially constructed during the reign of the Yongle Emperor in the Ming Dynasty (AD 1403–1424), but reduced in the 18th year of the reign of the Qianlong Emperor (1753)		

Sources: *Ming Hui Dian* (Administrative Laws of the Ming Dynasty), The Transportation Route in the Area North of the Yangtze River.

Chapter 3

The North Canal

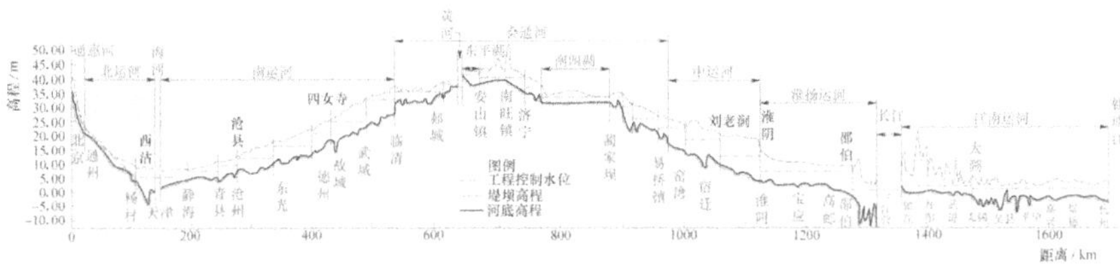

Figure 3.1: The North Canal

3.1 Introduction

Connecting with the Tonghui Canal and the South Canal in the north and south respectively, the North Canal (see Figure 3.1) was the northern end section of the Beijing–Hangzhou Grand Canal. Starting from Tongzhou, Beijing, and entering the Haihe River through the Sanchakou entrance in Tianjian, the canal was 148 km long. The North Canal was known as the 'Gushui River' before the Western Han Dynasty, the 'Lushui River' between the Eastern Han and Eastern Jin dynasties, the 'Luhe River' during the Northern Dynasty, and 'Baihe River' during the Yuan Dynasty. The Northern Canal Bureau was set up in the 9th year of the reign of the Shunzhi Emperor in the Qing Dynasty (AD 1652), and the canal has been known as the 'North Canal' since then.

In the 5th year of the reign of the Zhangzong Emperor in the Jin Dynasty (AD 1205), the Lushui River was made available for water transport, which utilised the downstream of the Baihe River by clearing the waterway. Thus, it became an important water transport route. Twelve rivers were connected at different times to form the North Canal water system (see Figure 3.2). In terms of ocean shipping, the North Canal was an important route connecting the ocean and inland rivers.

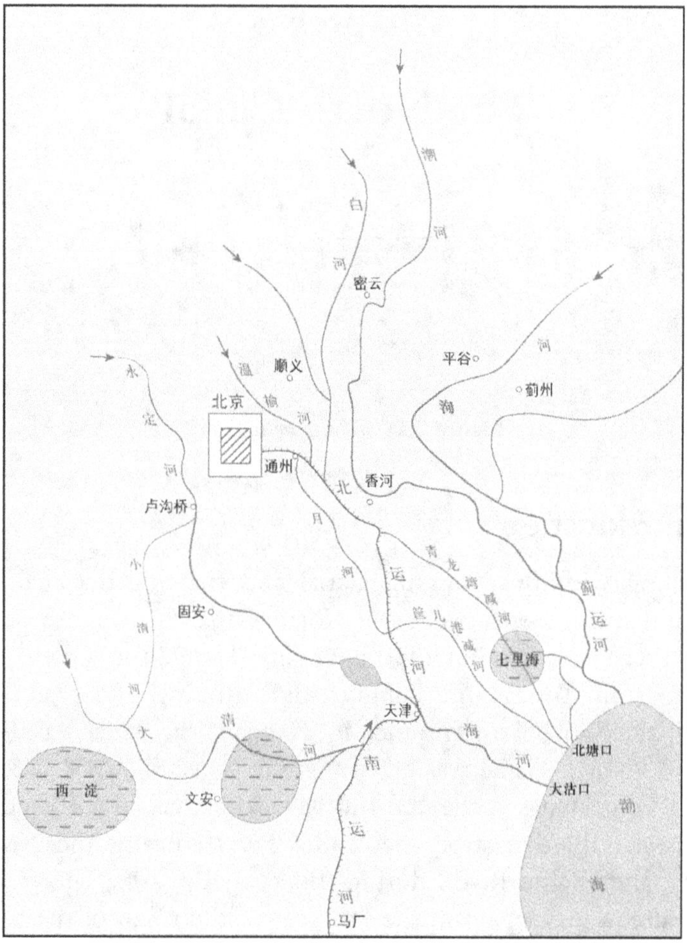

Figure 3.2: The North Canal Water System Between the 17th Century and the Middle of the 20th Century

Notes: The upper water source of the North Canal, the Wenyu River, rose from the south foot of the Yan Mountains in Changping, Beijing. Water from the north gate of Tongzhou was called the 'North Canal'. It reached Tumenlou Tower, passed the Qinglongwan water-discharge channel and joined the Baixin New River. Upstream of the Chaobai River were the Chaohe River and the Baihe River. Both of them rose in the south of Guyuan County, Hebei Province, and then joined together near Miyun; the river was called the 'Chaobai River'. The river reached the NaHuaihe River in Rouhuai and entered the plains. The downstream passed Suzhuang and reached the Xianghe River. The Chaobai New River went past Huangzhuangwa and Qilihai before joining the Yongding New River at Ningchegu and entering the ocean.[1]

[1] Le Huaichun, *300 Questions of Oceans and Rivers*. Zhengzhou, Yellow River Conservancy Publishing, 1999: 15.

3.2 Background

3.2.1 *Natural Conditions*

The main stream of the North Canal, with a total length of 148 km, starts from the check gate in the north of Tongzhou, Beijing, flows to the southeast through Xianghe County, Hebei Province, to Wuqing County, Tianjin, and joins the Xihe River (Ziya River) at the Xinhong Bridge in Tianjin.

The upstream of the North Canal is surrounded by mountains and hills that belong to the Yan Mountains. The general elevations of the canal are between 500 and 1,000 m, with the highest mountain peak at 1,000 m. The midstream and downstream of the canal are in plains with waterlogged areas. The basin covers an area of 6,166 km^2, among which the mountain and hill areas make up 952 km^2 and the plain area make up 5,214 km^2.

The North Canal basin is in the continental monsoon zone. The annual mean rainfall here is 643 mm. Most of the rainfall occurs between June and September, taking up 84 per cent of the annual total. Based on data from 1956 to 2000, the mean annual runoff was 4.81×10^8 m^3, of which 1.29×10^8 m^3 and 3.52×10^8 m^3 were from the mountains and the plains, respectively.

The rivers ran through the mountain areas, becoming the water sources for the upstream of the North Canal; therefore, the flow of the water was rapid and contained a large amount of sand. Riverbeds with fine silt soils resulted in severe sedimentation problems. Dikes were built on both sides of the canal to contain the water. Spacing between dikes varied from 100 m to 2,000 m. During flood seasons, floods often could not be discharged smoothly in the channels, meaning that the dikes burst with disastrous results. From the 6th year of the reign of the Chenghua Emperor in the Ming Dynasty (AD 1470) to the 7th year of the Republic of China (1918), a span of 449 years, 19 superfloods were recorded (alongside countless smaller floods). From 1918 (when hydrological observation data became available) to 1948, there were four superfloods. A particularly bad one occurred in 1939, when the floods of the North Canal and the Chaobai River joined together and resulted in catastrophe.[2]

---

[2] *The Beijing Local Chronicles Compilation Committee* (*Beijing Chronicles: Water Resources Chronicles*), Beijing Publishing House, 2000: 166–167.

With the development of railways and highways, the water transportation function of the North Canal gradually became superfluous from the end of the Qing Dynasty. Currently, the North Canal is mainly used for flood discharge, irrigation, and water transmission.

3.2.2 *History*

The current North Canal was mainly formed during the Jin Dynasty. After the Tonghui Canal was constructed during the Yuan Dynasty, the North Canal (see Figure 3.3) became a part of Beijing–Hangzhou Grand Canal and was included in the Grand Canal management system. In 1912, the Chaobai River changed course into the Ji Canal. Due to a shortage of water, water transport along the North Canal was abandoned.

In the 700 years from the beginning of the Jin Dynasty to the end of the Qing Dynasty, water transport sometimes combined river and ocean shipping. Whether it was by river or ocean, the North Canal was the essential path for the shipping of tribute rice. The North Canal played a very important role in the Beijing–Hangzhou Grand Canal system.

Huitong River was dredged in the 20th year of the reign of the Zhiyuan Emperor during the Yuan Dynasty (AD 1283). However, due to a water shortage from the north of Jining to Dongping, Shandong Province, the majority of water transport was achieved through ocean shipping. Ships travelled to Zhigu (modern-day Tianjin) and then went along the North Canal and the Tonghui Canal to reach the capital.

The North Canal's main project during the Yuan Dynasty was channel improvement, although there were also some early dikes and straightening projects.

With its improvement in the Jin and Yuan dynasties, the North Canal became the main stream in the north Haihe system. The canal was wide and shallow, with a high sediment concentration, and the amount of runoff varied widely between the flood and dry seasons. In winter and spring, with smaller volumes of water, the canal dried up. Water rose in summer and autumn, with the risk of bursts and

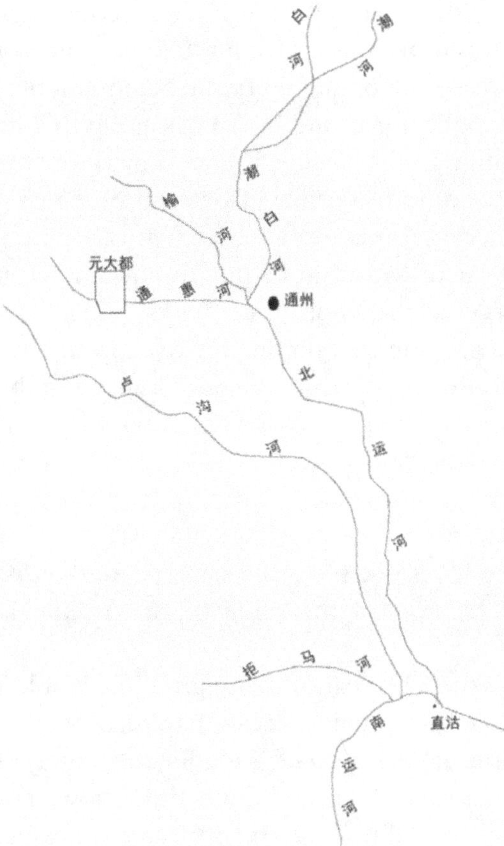

Figure 3.3: Route Map of the North Canal in the Yuan Dynasty

Notes: The Lu River was silted up at the end of the Jin Dynasty. At the beginning of the Yuan Dynasty, ships were unable to travel between Tianjin and the capital. Once the capital was built in the Yuan Dynasty, water sources were dredged to fill the canal. During the Yuan Dynasty, the Baihe River rose in north Tongzhou. It flowed south through Tongzhou, southeast to the boundary of Xianghe County, and entered Wuqing County until it reached the boundary of Jinghai County. During the Yuan Dynasty, all the ships from the ocean or the southern inland rivers and the ships from foreign seas had to travel along the North Canal to reach the capital. By that time, the North Canal was already very important.

diversions. It was recorded that the 'bursts and diversions [were] quite the same as the Yellow River's'.[3] The section between Hexiwu and Shua'erdu was notorious as being the most dangerous part of the North Canal. From the Yongle Era to the beginning of the Chenghua Era, there were eight bursts and blockages along this stretch.

At the beginning of the Ming Dynasty, maritime prohibition was implemented in southern coasts. However, tribute rice was still delivered through ocean routes under the control of the government. From the 9th year of the reign of the Yongle Emperor (AD 1411), the Emperor ordered the Minister of Works, Song Li, to reconstruct the Huitong River. Song Li dredged a river from the south of the Yangtze River to Beijing. Ocean transport stopped in the 13th year of the Yongle Emperor's reign (AD 1415), after which all water transport was undertaken via rivers.

After Beijing was chosen as the capital in 9th year of the reign of the Yongle Emperor in the Ming Dynasty (AD 1411), the construction of the Baihe River dikes was accelerated. All the dikes were completed in the 21st year of the reign of the Yongle Emperor (AD 1423).

In August of the 31st year of the reign of the Wanli Emperor (AD 1603), the Ming government decided to dredge the Baihe River entirely, requiring the depth to be 4 chi 5 cun (1.5 m). The dredging project began between April and June the following year and was commanded by Pan Dafu. Once the project was finished, their work was tested by a flood season: 'In August and September, water went down. Ships came from the south without any obstacles'.[4] This was the biggest dredging of the latter part of the Ming Dynasty.

The North Canal's engineering management focused on dredging the waterway. During the Yongle Era (1403–1424), River Control Governor Chen Xuan used to 'set 568 departments with solders to guide [the] ships to avoid shallow parts' along the canal between

[3] *Ming Dynasty History: River Chronicles IV*, Explanations on River Chronicles of 25 Histories. Zhou Kuiyi *et al.* Beijing, China Bookstore Publishing House, 1990:422.

[4] *Ming Dynasty Records: Emperor Shenzong Records, Book 407*. Research Institute of History and Language of Academia Sinica (ed.): 7597.

Huaisang and Tongzhou.[5] 'The Department' was the basic waterway management institute for guiding ships on the canal.

In the 3rd year of the reign of the Zhengtong Emperor (AD 1438), many *Qianpu* (departments for boat trackers) were set up along the canal between Hexiwu and Tianjin. Each *Qianpu* had several boat trackers. Their job was to clear the waterway during annual maintenance, build dikes, and pull ships when the canal was running.[6] In the 46th year of the reign of the Wanli Emperor in the Ming Dynasty (AD 1618), there were 59 *Qianpu* along the Baihe River between Tianjin and Tongzhou. Each year, all the cities, counties, and states would send over 1,700 workers. In the 9th year of the reign of the Hongzhi Emperor (AD 1496), there were 65 *Qianpu* on the Baihe River.[7] In the 28th year of the reign of the Jiajing Emperor (AD 1594), the Ming government issued an order to clear the waterway north of Tianjin in March each year. The river managers were told to go to the *Qianpu* themselves and command the work. That same year, some water observatories were set up between Hexiwu and Tongzhou to measure the canal's water volume.[8]

During the Qing Dynasty, the North Canal was not only used as a water transport channel, but also as an important flood control project. In the 3rd year of the reign of the Yongzheng Emperor (AD 1725), the Baihe River was flooded. Many dikes were burst. Emperor Yongzheng ordered Prince Yunxiang to control the water and build water-discharge dams and channels at dangerous parts of the canal. Water was discharged and diverted through different routes into the ocean. The rivers in the capital were divided into four systems, and the North Canal was one of them. Since then, the 'North Canal' replaced the old names of the 'Luhe River' and the 'Baihe River'.

---

[5] *Ming Dynasty History: 41st Biography: Chen Xuan.* (Qing Dynasty) Zhang Tingyu. Beijing, Zhonghua Book Company, 1974: 4208.

[6] 'Record of Laws and Systems of Ming Dynasty' from *Water Conservancy History, Book 104* (Qing Dynasty) Fu Zehong. Beijing, Commercial Press, 1951: 1528.

[7] *Graphical Chronicles of River Transport*, Yao Hanyuan, Tan Xuming. China Water Power Press, 1990: 16–23.

[8] *Ming Dynasty Records: Emperor Shenzong Records Book 350*, Research Institute of History and Language of Academia Sinica (ed.): 6327–6328.

Figure 3.4: Grain Shipping on the Luhe River in the Qing Dynasty

The flood control projects carried out during the Qing Dynasty were Kuang'erwan, Hexiwu, Qinglongwan, and other water-discharge dams and channels.

Due to the disadvantages of river transport and the success of attracting ocean shipping, in the 27th year of the reign of the Guangxu Emperor (AD 1901), the Qing government issued an order to stop river transport.[9] As the Jingshan Railway provided a more convenient path for the delivery of food, the transportation of grain (see Figure 3.4) along the North Canal — which had lasted for hundreds of years — finally came to an end. This was in spite of the fact that Jiangsu and Zhejiang had to collect millions of kilos of tribute rice to balance the food price in the capital.

3.3 Key Projects

The North Canal was established from the Chaobai River (which was called the 'Luhe River' below Tongzhou). It had a high gradient and

[9] *Qing Dynasty History: River Chronicles II*, Explanations on River Chronicles of 25 Histories. Zhou Kuiyi *et al.* Beijing, China Bookstore Publishing House, 1990: 587.

variation of discharge in high and low water volumes. Downstream, the rivers joined together and became the main stream of the Haihe River. Northeast of the Haihe River's main stream was the Sanjiaodian Lake. Many flood control projects had been built here since the Ming and Qing dynasties, and so this became the flood detention region to protect Tianjin.

Though the North Canal didn't have many facilities, its curves and water-discharge dams reduced the waterway's vertical gradient and controlled the water level. It was able to support water transport and discharge floodwater. Water-discharge dams were built at the joint of the canal and the natural rivers. Next to the dams were the water-discharge channels (also called *Yinhe* channels). They could divert floodwater to enter the lakes along the downstream of the Yongding River or into the Bohai Sea in the east. The main engineering effects of water-discharge dams were to control the volume of water in the canal, prevent the bursting of dikes, and guarantee safe transport along it. Famous examples of water-discharge projects on the North Canal were the Hexiwu–Qinglongwan Junction and the Kuang'erwan Junction (see Figure 3.5).

The allocation of the water-discharge dams effectively reduced the risks of the dikes bursting. Sections with these dams formed engineering junctions to control water volume and connect natural rivers with artificial waterways. In terms of controlling floods in urban areas, these water-discharge channels diverted part of the floodwater to both the South and North Canals, meaning that the Tianjian Sanchakou area had to bear much less floodwater. There was only 60 km of sectional floodwater on the North Canal. In the 14th year of the reign of the Qianlong Emperor (AD 1749), Fang Guancheng was appointed as the Governor of Zhili. He commented on the water-discharge channel's dredging and clearing, stating, 'They can protect not only water transport and dikes, but also the Sanchakou area.'[10]

[10] *4 Cases of Capital Water Junction: Case Four* (Qing Dynasty), Pan Xi'en, Chinese Institute of Water Resources and Hydropower Research: 67.

Key:

香河县—Xianghe County　　王家务减水坝—Wangjiawu Water-Discharge Dam
河西务Hexiwu　　青龙湾引河—Qinglongwan Water-Discharge Channel
河西务引河—Hexiwu Water-Discharge Channel
筐儿港减水坝—Kuang'ergang Water-Discharge Dam
筐儿港引河—Kuang'ergang Water-Discharge Channel
同治十三年新开—Opened in the 13th year of the reign of the Tongzhi Emperor
蓟运河—Ji Canal　　北运河—North Canal
永定河—Yongding River　　七里海—Qilihai Lake
子牙河—Ziya River　　南运河—South Canal
天津城—Tianjin　　海河—Haihe River
渤海—Bohai Sea　　减水坝—Water-Discharge Dam

Figure 3.5: An Illustration of the North Canal's Important Water-Discharge Junctions and Water System

3.3.1 *Kuang'ergang Junction*

Kuang'ergang Junction is located in modern-day Wuqing County, Tianjin. Between the 36th and 38th years of the Kangxi Emperor's reign (AD 1697–1699), many bursts occurred on the North Canal's Kuang'ergang section in Wuqing County. The transport route was blocked. In the 39th year of his reign (AD 1700), Emperor Kangxi went to see the bursts for himself, investigating and deciding to build stone water-discharge dams at the bursts in addition to some water-discharge channels.[11] In the 43rd year of his reign (AD 1704), the Kuang'ergang water-discharge channel was dredged. The dam was 600 m wide. Water flowed in channels around Yangcun Village along a radius of approximately 50,000 m, and the dikes were substantial. Emperor Kangxi wrote 'divert water and support transport' on the Kuang'ergang water-discharge dam monument (Figure 3.6).[12]

Figure 3.6: The 'Divert Water and Support Transport' Monument by Emperor Kangxi (photo taken in April 1984)

Note: The original monument still stands near the Kuang'ergang Junction.

---

[11] *Qing Dynasty History: Chronicle 27: Geography I*, Zhonghua Book Company, 1977: 1897.

[12] *Capital Anlan Chronicles: Beihe Book IV* (Qing Dynasty), Wang Lyutai, Chinese Institute of Water Resources and Hydropower Research: 11.

Figure 3.7: The Kuang'ergang Water-Discharge Channel in 2010

Afterwards, there was water clearing, widening, and deepening of the river.[13] The Kuang'ergang River currently runs past the Jinzhong River and enters the ocean (see Figure 3.7).

The Kuang'ergang water-discharge channel (see Figure 3.8) began in Wuqing District, on the left bank of the North Canal, and flew east. Then it turned south and passed Zhujia Port, Meichang, Guojiatai, Caijiazhuang, and the Yangjia River before passing Hanshengzhuang and entering Maizidian Lake; it then entered Tahedian Lake through the Yaohe River, ran past the dike and exit in the east, and then went past onto the end of the west dike, exiting through Qilihai Lake in Ninghe County in the east and entering the

[13] *Capital Anlan Chronicles: Beihe Book III*: 19.

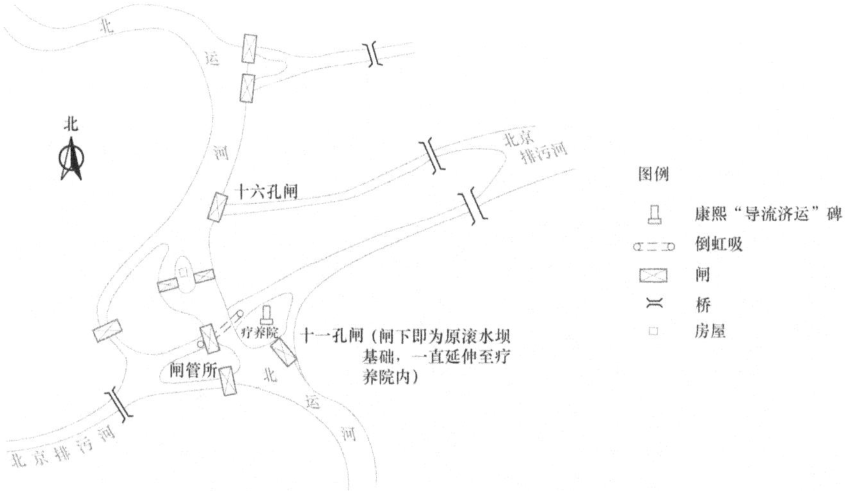

Key:

北运河—North Canal　　北—North

十六孔闸—16-Hole Gate　　北京排污河—Beijing Sewage Discharge Channel

闸管所—Gate Management Institute　　疗养院—Sanatorium

十一孔闸—11-Hole Gate

康熙'导流济运'碑—Kangxi 'divert water and support transport' monument

倒虹吸—Inverted Siphon　　闸—Gate　　桥—Bridge

Figure 3.8: The Location of the Remains of the Kuang'ergang Water-Discharge Channel's Stone Dam

Notes: In the 1930s, the Kuang'ergang water-discharge channel's stone dam was changed to an eight-hole gate. In 1960, it was changed into an 11-hole gate (see Figure 3.9). Until now, the base wooden stakes and stones of the original Gunshui Dam can still be found to the front and rear of the dam. Currently, the gate sits over the Beijing Sewage Discharge Channel, a part of the Kuang'ergang Junction. It is used to discharge floodwater during flood seasons. Compared to the old Kuang'ergang water-discharge channel, the Beijing Sewage Discharge Channel's path has changed greatly.

Ji Canal. It finally joined the ocean through Beitang. During flood seasons, floodwater would bring mud and sands into Tahedian Lake, thus reducing the siltation in the North Canal. There was less water in the springs, and so the water-discharge channel would be closed to guarantee the water level in the North Canal would not drop below that needed to facilitate water transport.

Figure 3.9: The Remains of the Kuang'ergang Water-Discharge Channel's Stone Dam (photo taken on the left bank of the downstream in April 1984)

Kuang'ergang Junction currently includes a new three-hole barrage, an old six-hole check gate, the Longfeng New River's six-hole check gate, and a sewage-discharge channel for inverted siphon.[14] The projects' main functions are flood control, flood regulation, water storage, irrigation, water supply, and pollution prevention and discharge.

3.3.2 *Hexiwu–Qinglongwan Junction*

Hexiwu–Qinglongwan Junction was an important flood control project on the North Canal. It was made up of the Wangjiawu and Qinglongwan water-discharge dams and the Hexiwu and Qinglongwan water-discharge channels.

[14] *Tianjin Water Resources Chronicles*. Tianjin Water Resources Bureau Water Resources Chronicles Compilation Committee. Tianjin: Science and Technology Press, 2003: 120.

The axis on the right bank of Wangjiawu Water-discharge dam

(a)

Masonry in the ruins of the water-discharge dam

(b)

Figure 3.10: The Ruins of the Wangjiawu Stone Water-Discharge Dam (1984)

Once the Kuang'ergang Junction was dredged, it diverted the pressure caused by the floods in the Yangcun area. However, it didn't help the upstream area, Hexiwu, at all. It was always a dangerous area around Hexiwu, with Shua'erdu suffering frequent catastrophic bursts. In the 8th year of the reign of the Yongzheng Emperor (AD 1730), a flash flood swept down the upstream of the North Canal. Hexiwu's Shua'erdu was burst. Emperor Yongzheng ordered the Governor of Rivers to build a 120-m stone water-discharge dam in the upstream north of Tumenlou Village (the junction of Wuqing and Xianghe counties) as well as a 45,000-m-long water-discharge channel (the modern-day Qinglongwan channel) in Wangjiawu and a long dike to discharge floodwater into Qilihai Lake. The project was completed in the 9th year of his reign (AD 1731).[15] The Qinglongwan River had a waterway clearance in 1972. Its path was changed at Badaogu Lake; it now went past Lizigu Lake and entered the Chaobai New River. However, its basic path did not change much. From Tumenlou Village, Xianghe County, in Hebei Province, it went past Xianghe County, Wuqing County, and Baodi County and entered the Chaobai New River. The stone water-discharge dam

[15] *Capital Anlan Chronicles: Baihe River Book III* (Qing Dynasty), Wang Lyutai, Chinese Institute of Water Resources and Hydropower Research: 19.

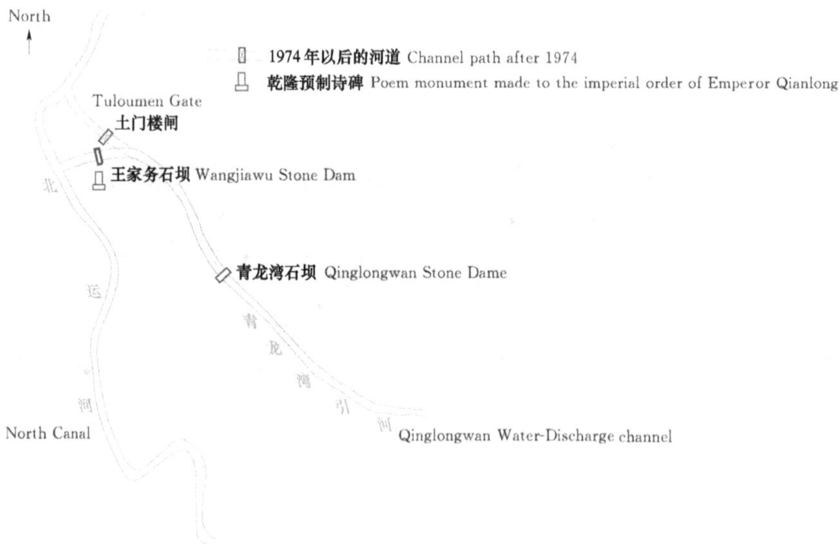

Figure 3.11: The Location of the Qinglongwan Stone Water-Discharge Channel (Based on Field Trip)

Ruins of rammed earth
(a)

Ruins of stone component
(b)

Figure 3.12: The Ruins of the Qinglongwan Stone Water-Discharge Dam (1984)

at the junction of the beginning of the Qinglongwan water-discharge channel and the main stream of the North Canal was modified into a flood-diversion gate in 1926, namely the Tuloumen Gate.

The original location of the Qinglongwan Stone Water-Discharge Dam (see Figures 3.11 and 3.12) was between Sanbaihu Village and

Wangjiawu Village in Xianghe County. Sanbaihu was on the left bank and Wangjiawu was on the right. Their distances to the dam were both around 500 m.

As the stone dam was in a low-lying area, the flood-discharge results after its completion were not good. In the 2nd year of the reign of the Qianlong Emperor (AD 1737), it was moved upstream by 2 km, closer to the upper entrance of the water-discharge channel. After this move, the flood-discharge path was unobstructed, allowing it to play an effective role during this period. It was recorded that 'water went over the dam entrance by over 3 m. [The water] was still calm. Thousands of lives were saved in Hexiwu'.[16] After the dam was moved, it was renamed as the 'Wangjiawu Water-Discharge Dam' (see Figure 3.10), with a length of 40 zhang (66.6 km).[17]

In the 37th year of the reign of the Qianlong Emperor, the Qing government lowered the top of the Wangjiawu water-discharge dam by 3 m to increase the volume of flood discharge.[18] Emperor Qianlong wrote a poem about the flood control effects of the project: 'By lowering the gate by 3 m, making it similar to a check gate, it cleared the waterway of the channel and protected the capital'.[19]

Based on the original dam, the Wangjiawu water-discharge dam was modified into a 40-hole floodgate in the 1920s. It was known as 'Tumenlou Gate' afterwards. In 1974, approximately 100 m north of the old gate, a new Tumenlou Gate was built. The Qinglongwan water-discharge channel's water intake was moved north. Soon afterwards, the old gate was blown up.

The Qinglongwan water-discharge channel is a first-grade seasonal channel managed by the city. It is 180 km long. The waterway was cleared and the dikes were mended in 1951 and 1972, respectively. The channel is 140 m wide. The channel's lowest elevation

[16] Quoted from *Qing Dynasty Haihe and Luanhe Flood History*, Chinese Institute of Water Resources and Hydropower Research: Item 75: 1737–1750.

[17] *Capital Chronicles: Book 87: Rivers 13* (Qing Dynasty) by Huang Pengnian, Chinese Institute of Water Resources and Hydropower Research: 3527.

[18] *Capital Anlan Chronicles: Baihe River* Book I & III: 11.

[19] (Republic of China) *Xianghe County Chronicles: Hongli Imperial Signed Jinmen Gate*.

is between 1.19 and 4.6 m; the tops of the dikes are between 9.15 and 15.54 m high. The distance between the two dikes is 280 m. The water-flow volume is 1,330 m^3/s. The watershed in Xianghe County is 36.3 km^2. With normal water flow, the channel can irrigate 3.996×10^7 m^2 of fields along the channel. Because the Chaozi River in Baodi County and the Qinglongwan River can store water, there is additional water that can be channelled. The channel can prevent flood, store water, supply water for irrigation, and accept water-discharge from the fields. It has 13 gates, culverts, and pumping stations in Xianghe County, as well as a railway and a highway bridge.

3.3.3 *Water-Conservancy Project*

Tianjin Sanchakou (see Figure 3.14) is the joining point of the South and North Canals, and also the starting point of the Haihe River. The joining point is currently near the Shizilin Bridge in downtown Tianjin, now called 'Old Sanchakou'. The area is relatively high in altitude, above 68 m on average (Dagu level). The threat of floods is therefore limited. Sanchakou was one of the earliest settlements in Tianjin. At the beginning, people made a living through fishing and salt-making, and people began to settle there. It was the earliest settlement in Tianjin since the middle of the Northern Song Dynasty.

During the Jin and Yuan dynasties, the Yellow River diverted south and stopped entering the ocean from Bohai Bay. Tianjin's coast and rivers were basically stable. Approximately 60 km from the Haihe entrance to Sanchakou and 20 km from Sanchakou to Yangcun are tidal reaches. They form an excellent route into the harbour for ships coming from the oceans and rivers. In the 4th year of the Jin Dynasty (AD 1204), Han Yu dredged a new transport channel, created a new river network with a terminal in Yanjing, and created 'Xin'an Ocean Shore' as the junction. From the structure of the channel, 'Xin'an Ocean Shore' is the entrance of the Sancha River in downtown Tianjin today. Tianjin's (known as 'Zhigu' in the Jin Dynasty) position as the throat of river transport to the capital was established. During the Jin Dynasty, Sanchakou had developed as an important base for the military and the communication network. Later in the Yuan, Ming, and Qing dynasties, river transport prospered rapidly here.

During the Jin Dynasty, the transport of tribute rice in Hebei, Henan, and Shandong all went through Zhigu Port to reach Zhongdu. In the 21st year of Jinshizong (AD 1181), over 600 kg of tribute rice was delivered to Zhongdu through Zhigu.[20] Zhigu Port played an important role as a transit port for food shipping.

During the Yuan Dynasty, Sanchakou became the intersection of ocean and river shipping. Ships entering the ocean and going into the Luhe River (now the North Canal) all gathered here.

It was a busy picture of 'tens of thousands of ships gathered at Sanchakou at dawn'.[21] Until the 22nd year of the reign of the Zhiyuan Emperor (AD 1285), under the governance of the 'Inter-River Salt Transport Governor', 22 salterns were established on the west coast of Bohai, Sanchagu, and Fengcai salterns now located in downtown Tianjin. Their annual output is 8×10^7 kg. Salt was sold all the way to Linqing, Tongzhou, and the capital. Zhigu also became a salt transport port.

During the Ming Dynasty, since ocean shipping was banned and food was all delivered through rivers, the delivery of salt climbed dramatically in the 15th year of the reign of the Yongle Emperor (AD 1417), going from over 5.6×10^8 kg in the previous year to over 1.016×10^9 kg. In the 7th year of the reign of the Xuande Emperor (AD 1432), the value reached a record of 1.348×10^9 kg. In the 8th year of the reign of the Chenghua Emperor, a rate was set at 8×10^8 kg. From then on, during the governance of Hongzhi, Zhengde, Jiajing, and Longqing, the delivery volume remained around this number.

Tianjin continued to be a river transport junction in the Qing Dynasty. With ocean shipping banned, Tianjin Port dealt with a transhipment of over 8×10^8 kg/year, which was the same as the

---

[20] *Jin Dynasty History: River Chronicles.* Explanations on River Chronicles of 25 Histories: 228.

[21] *Tianjin County Chronicles: Book 22: Sacrifice to Tianfei Temple, Made in Zhigu* (Yuan Dynasty) by Zhang Zhu: 2.

Figure 3.13: Ships Waiting to Pass the Gate in the Canal near Tianjin, 1938

Figure 3.14: Tianjin Sanchakou in 2013

Note: Currently, Tianjian Sanchakou takes up an area of approximately 1 km^2.

Ming Dynasty. Due to the North Canal's worsening siltation, access to the canal was blocked. People had to use barges to tranship from the Tianjin Port (see Figure 3.13). At the beginning of the Qing Dynasty, there were 600 red barges; this grew to 1,500 in the 60th year of the reign of the Qianlong Emperor (AD 1795). After the river ships arrived at Tianjin for the first time, 6×10^7 kg had to be unloaded and stored in the county's storage temple. The rest would be delivered to Tongzhou with barges. Food stored in the Tianjin storage temple would then be delivered to Tongzhou in batches.

After ocean shipping was allowed in the Daoguang years, ships entering Tianjin Port from the ocean had to go through the transhipment procedures in Gagu. Tianjin Port's transhipping ports were scattered on both banks of the Haihe River, around the new river entrance, Gagu, outside the East Gate, and at the northeast corner. Along the North Canal, Dingzigu, Beicang, Yangcun, Caicun, Hexiwu, and Yangliuqing and Beidaguan along the South Canal were famous transhipping ports.

Chapter 4

The South Canal

4.1 Introduction

Starting from the Linqing hub in Shandong Province and connecting with the Huitong River, the South Canal went from south to north through Dezhou in Shandong, Dongguang, Botou, Cangxian, and Qingxian in Hebei Province, to Jinghai County in Tianjin, and then connected with the North Canal. The canal was 446 km long with a longitudinal slope from south to north of 0. 03 per cent (see Figure 4.1).

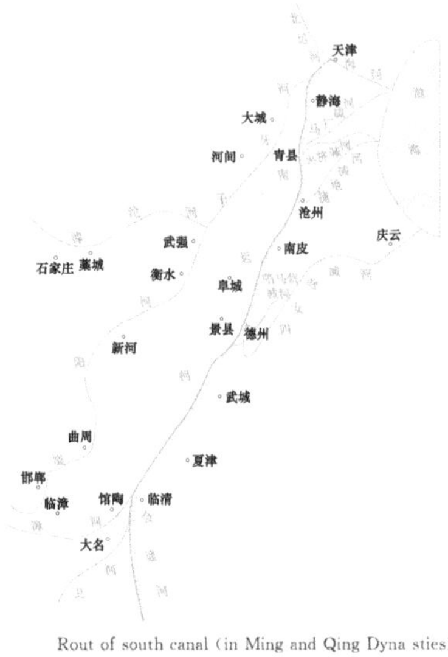

Rout of south canal (in Ming and Qing Dyna sties)

(a)

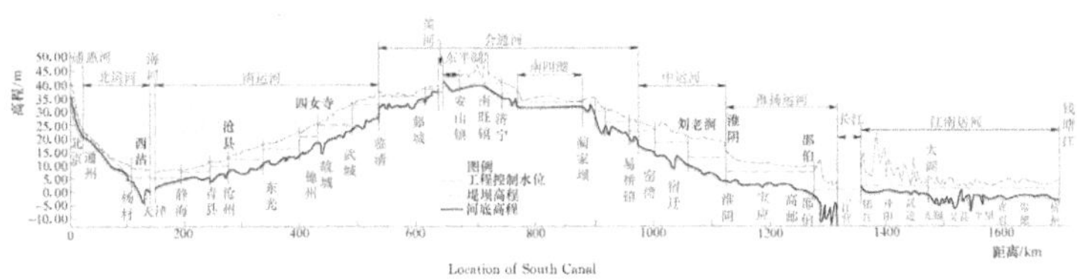

Location of South Canal

(b)

Figure 4.1: Flow Route and Location of the South Canal

4.2 Natural Conditions

The South Canal — known as the Weihe River on the upper reaches of Shandong Linqing — originated from the east foothills of the Taihang Mountains within the territory of Shanxi and merged mainly with the Zhanghe River, TuHaihe River, and Ziya River along the way. The average yearly precipitation over the South Canal basin was 550–800 mm. Most of the rainfall occurred in the summer, with the precipitation in July and August accounting for 60 per cent of the annual total. This caused great variation in the water flow of the canal during the dry and wet seasons.

The main characteristics of the South Canal's water sources included the small total amount, extreme variations according to space and time, and frequent occurrence of sequential droughts. The amount of water available per person in the region was only 240 cubic metres, less than one ninth of the national level. Therefore, this region was severely short of water resources.

Since the 1980s, the South Canal has frequently been without water. Due to the increased exploitation of groundwater, depression funnels were formed in Dezhou and Cangzhou within the middle and lower reaches of the South Canal basin. The aquatic environment of the South Canal has seriously deteriorated.

4.3 Historical Development

In the late Eastern Han Dynasty, Cao Cao (the Leader of the Wei Country) ordered the excavation of Baigou, the Pinglu Canal, and the Licao Canal to interconnect all the water areas of the Weihe River. During the Sui Dynasty, the Yongji Canal was excavated to form a main waterway to connect with the North China Plain. The Weihe River was often called the 'Yuhe River' in the Tang, Song, and Yuan dynasties. The Beijing–Hangzhou Grand Canal was completed in the Yuan Dynasty and included sections of the Yuhe River in the north of Shandong Linqin. During the Ming and Qing dynasties, the section of the Beijing–Hangzhou Canal from Linqing in Shandong to Tongxian in Beijing was known as the 'Beihe River', which was divided into Weicao and Baicao according to different water sources. During the Qing Dynasty, the section of the Beijing–Hangzhou Canal from Linqing in Shandong to Tianjin was known as the South Canal.[1]

4.3.1 *The Yuhe River in the Yuan Dynasty*

After the successful excavation of the Huitong Canal in the 26th year of the reign of the Zhiyuan Dynasty in the Yuan Dynasty (AD 1289), the Yuhe River merged with the Huitong Canal in Linqing, Shandong. Boats carrying tribute grains could now go up from the Yellow River to Xuzhou, then to the Huitong Canal, entering the Yuhe River through Linqing and then finally reaching Beijing. Since then, the downstream section of the Yuhe River from Linqing became known as the South Canal of Beijing–Hangzhou Canal system, and the whole route of the Beijing–Hangzhou Canal was approximately 1,000 km shorter than the Sui–Tang Grand Canal.

The South Canal was the main stream of southern system of the Haihe River. The major water transportation problems included great variations in the water level during the dry and wet seasons, frequent embankment breakage in the flood seasons, and insufficient water in the dry seasons. Therefore, in the Yuan Dynasty, two types of engineering measures were adopted. The first was the construction of flood-reducing rivers, and the second was the utilisation of the irrigation water from upstream of the Weihe River to guarantee a water source for the canal during the dry months of winter and spring. These two methods continued to be used until the Qing Dynasty.

4.3.2 *The South Canal in the Ming and Qing Dynasties*

In the early years of the reign of the Yongle Emperor in the Ming Dynasty, a proportion of the tribute grain from the lower reaches of the Yangtze River was transported over the ocean, entering the ocean

---

[1] (Qing Dynasty) Li Shixu: *Continuous Gold Guidance for Water Flowing*, Vol. 67, Shanghai: Commercial Press, 1938.

from the river. This meant the vessels entered the Baihe River after leaving Zhigukou; the rest of the grain, which was transported by water and by hand, entered Huaihe from the river, went up to Yangwu along the Yellow River, reached Weihui after 85 km of road transportation, and finally entered the Weihe River.

In 13th year of the reign of the Yongle Emperor (AD 1415), the Count of Pingjiang, Chen Xuan, excavated the Huitong Canal again so that the ships could enter the Weihe River in Linqing. At that time, it was called 'Weicao'. Thus far, ocean transportation and land transportation were basically stopped. The Weihe River only saw the transportation of tribute grain from the nearby provinces of Shandong and Henan, starting in a new place — Xiaotan Town, Damingfu — and reaching Shandong Guanxian after travelling 30 km towards the northeast. In the Ming and Qing dynasties, the Huitong Canal became a main transportation route, enabling the connection of the Huanghai Sea and the Huaihai Sea, but suffered frequent embankment breakage in the flood season, insufficient water in the dry season, and so on. Therefore, the measures of leading the Zhanghe River and the Qinhe River into the Weihe River were adopted to support transportation, with the intentional excavation of new rivers to discharge floodwater.

4.3.2.1 *Transportation-Supporting Works*

During the Ming and Qing dynasties, the Weihe River frequently suffered a shortage of water at its upstream source during the dry season, so the availability of water was increased through measures such as drawing water from other rivers and controlling the upstream irrigation water of the Weihe River. Transportation-supporting rivers converging on the upper reaches of the Weihe River mainly included the Danhe River, the Qinhe River, the Qihe River, the Zhanghe River and others; the Zhanghe River and Qinghe River were particularly useful in this regard.

After the Yongji Canal was excavated during the Sui Dynasty, Zhanghe began to become one of the most important water sources for the South Canal. After this, the route of the Zhanghe River was

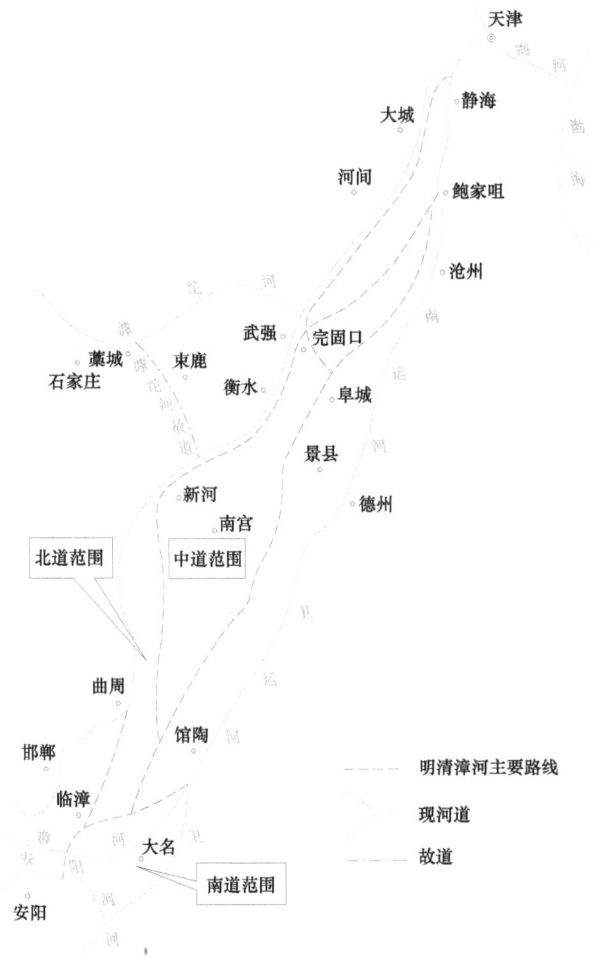

Figure 4.2: Diagram of the Changing Course of the Zhange River in the Ming and Qing Dynasties

frequently changed (see Figure 4.2), while the water source was still unstable. In the 13th year of the reign of the Zhengtong Emperor in the Ming Dynasty (AD 1448), Supervisory Censor Lin Tingju presented an application to arrange a sluice gate on the Zhanghe River to lead water flow into the Weihe River; as such, the damage caused by the water of the Zhanghe was prevented and transportation along the Weihe was supported. This provided an important precedent in terms of introducing water from the Zhanghe River into the Weihe River on a large scale.

The Baiquan Spring, as the source of the Weihe River, had been used for irrigation since ancient times; in the early years of Ming Dynasty, five gates — Ren, Yi, Li, Zhi, and Xin — were established and regularly opened or closed to support not only transportation, but also farmland irrigation. During the Ming and Qing dynasties, supporting transportation predominated; if there was surplus water (except during canal transportation season), then irrigation could be allowed. Besides the Weihe River, the Qing government also controlled the irrigation water of rivers flowing into the Weihe River (and particularly the Xiaodanhe River).

4.3.2.2 *The Water-Reducing River*

The South Canal, on the lower reaches of the Weihe River, was also a flood-discharge channel of the Weihe River. Since the Zhanghe River was changed to enter the Weihe River from Guantao, the capacity for floods was increased, and it was recorded that '[the] narrow waterway and lower location easily result[ed] in embankment breaks; and in the occurrence of each embankment breakage, a plurality of workers [were] allocated to repair'. Therefore, since the Yongle Era of the Ming Dynasty, multiple water-reducing rivers were successively excavated along the east side of the North Canal in Dezhou to discharge the flood into the ocean. From south to north, these excavated water-reducing rivers were Sinyusi in Dezhou, Shaomaying, Jiedi in Cangzhou, Xingji (between Cangzhou and Qingxian County), and Tianjin Machang (see Figure 4.3). Water gates and dams were constructed for each river mouth to control the amount of water discharged.

4.3.2.3 *Desilting and Embankment Reinforcement*

During the Qing Dynasty, systematic embankments had been built for the South Canal from the downstream of Linqing. The high sediment concentration in the Zhanghe River resulted in an accumulation

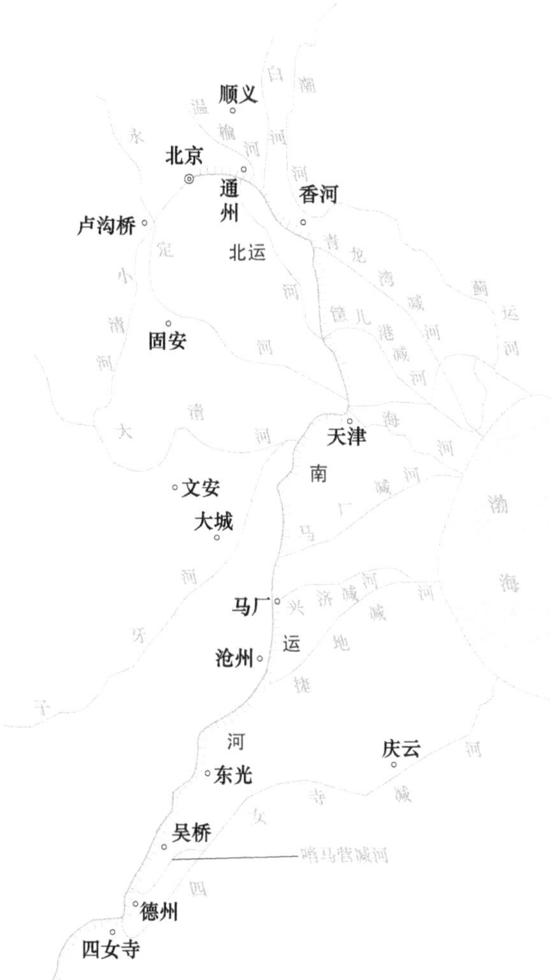

Figure 4.3: Distribution Diagram for Water-Reducing Rivers Along the South Canal and North Canal (C. 15th–19th)

of sand in the South Canal year-on-year. Not only were the urban areas and rural fields and houses located on lower terrain, but the river bends also stagnated; consequently, many sections of the canal were dangerous. Therefore, Emperor Qianlong adopted the measures of reinforcing crescent dikes and flattening dangerous stagnating portions of the South Canal. Under the leadership of Tianjin Director

Chen Hongmou, the South Canal was successfully desilted at more than 30 locations from the 3rd year of the reign of the Qianlong Emperor to the 5th year of his reign (AD 1738–1740).[2] The desilting and embankment reinforcement measures on the South Canal were mainly implemented during the reign of the Qianlong Emperor in the Qing Dynasty.

4.4 Critical Projects and Scientific Value

The Zhanghe River and the Hutuo River had a high sediment concentration and converged into the upper reaches of the South Canal, which was the trunk stream of the southern system of the Haihe River. On the other hand, the differential water level, the terrain gradient, and so on were caused by the extremely great high–low water flow of the Weihe River, which was unhelpful in terms of water transport. During the Yuan, Ming and Qing dynasties, the South Canal was renovated through the intentional prolonging of the bends to decrease the longitudinal slope of the waterway and slow the flow. Not one gate was constructed, but the adjustment of the channel's hydraulic characteristics and the canal water levels was successful, generating the mantra that 'three bends are equivalent to one gate'. To meet the demands of discharging floodwater through the trunk stream in the flood season, controlling the amount of water on the waterway, and guaranteeing smooth transport along the canal, the Sinyusi, Shaomaying, Jiedi, Xingji, and Machang water-reducing rivers and the like have been successively excavated at dangerous locations along South Canal since the Ming Dynasty. The water-reducing dams were arranged at the intersections between each water-reducing river and the canal, and the flood was split through the water-reducing dams and discharged into the ocean through the water-reducing rivers.

Water-reducing gates with dams (masonry multi-hole water gates) were arranged along the right bank of the canal as main work

---

[2] (Qing Dynasty) Li Shixu: *Continuous Gold Guidance for Water Flowing-Volume 95. Basic Books for Studies of Chinese Traditional Culture*. Shanghai: Commercial Press, 1938: 2153.

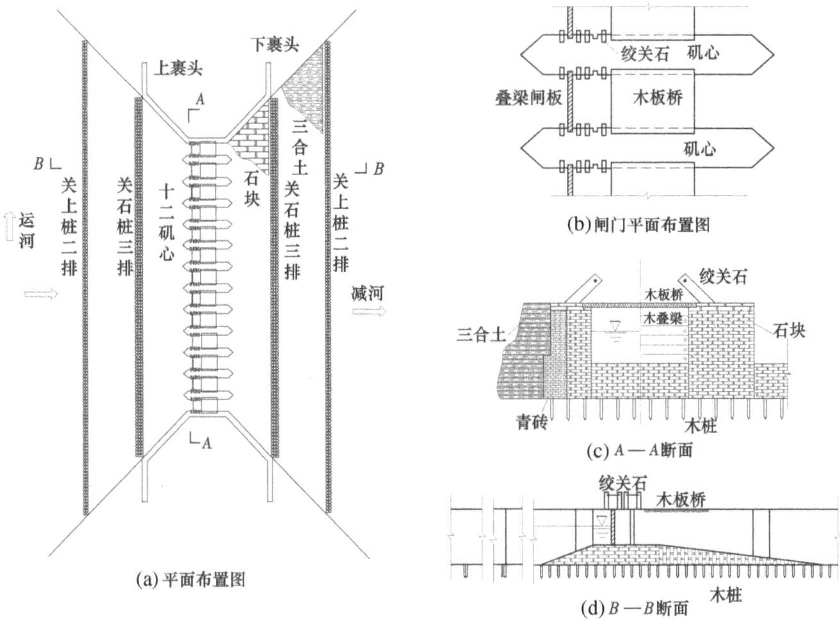

Figure 4.4: Recovery Plan of Water-Reducing Gates with Dams During the Qing Dynasty

appliances for controlling the amount of water in the South Canal. At present, the Jiedi Gate and Jiuxuan Gate (reconstructed from the Machang water-reducing gate) are still standing. Figure 4.4 shows a recovery plan of water-reducing works on the South Canal, drawn according the records of documents such as 'Flood-Prevention Work Minutes' (compiled by Zhou Jiajyu in the Qing Dynasty), 'Gold Guidance for Water Flowing', and 'Continuous Gold Guidance for Water Flowing', in addition to investigation on existing similar works.

4.4.1 *Sinyusi Junction*

The construction of Sinyusi Junction (see Figures 4.5 and 4.6) was initially designed as a water-reducing work. In the 9th year of the reign of the Yongle Emperor in the Ming Dynasty (AD 1411), the Huitong Canal was excavated again. Due to the overflow of the Weihe River in this year, Minister of Works Song Li implemented the

Figure 4.5: The Scene of the Sinyusi Hub (provided by the Zhanghe River, Weihe River, and South Canal Authority)

Figure 4.6: The Old Ship Lock in Sinyusi (2009)

introduction of the Wen River into the Weihe River in the north through the Huitong Canal and suggested excavating a total of 228.5 km for a new river — the predecessor of the Sinyusi water-reducing river — in the northwest of Dezhou City. This would split the discharged water into the old route of the Yellow River and then into the ocean in the port of the Dagu River. The idea was that the rising Weihe River would flow back to the Huitong Canal. After a successful excavation, the overflowing water of the Zhanghe River flowed into the Fuyang River in the north, resulting in a gradual stagnancy of the waterway leading east to the Weihe River; as a result, no function was exerted.

In the 2nd year of the reign of the Hongzhi Emperor (AD 1489), the embankment of the Yellow River was broken in Kaifeng and Jinlong Port and the water flowed into the Zhangqiu Canal in the north. Minister of Revenue Bai Ang implemented the excavation of 12 rivers from Dongping to Xingji to divert the flood into the ocean through the Qinghe River and the old route of the Yellow River. He also oversaw the construction of a weir that was opened or closed quickly according to whether the canal was rising or falling. In the 10th year of the reign of the Jiajing Emperor (AD 1531), the Zhili Province Censor Zhan Kuan suggested constructing a stone gate in Dezhou to let the flood flow into the ocean. Four years later, the Dezhou Sinyusi water-reducing gate and four other such gates in Jingzhou Botou Town were recovered to divert the flood into the ocean. Thirty-seven years later, in the 1st year of the reign of the Wanli Emperor (AD 1573), the embankment of the Zhanghe River was broken in the north, while the flood flowed into the Fuyang River — not the Weihe River. Therefore, all the gates were gradually abandoned.

In the early years of the Qing Dynasty, the Sinyusi water-reducing river was silted up and the gate base was damaged, so Shandong and Zhili provinces frequently suffered from the canal overflowing. In the 44th year of the reign of the Kangxi Emperor (AD 1705), the water-reducing gate was recovered. In the 4th year of the reign of the Yongzheng Emperor (AD 1726), Secretary of the Grand Secretariat He Guozong presented a proposal to change the Sinyusi water-reducing gate into an overflow dam. The Sinyusi water-reducing river underwent maintenance several times during the Ming and Qing dynasties, mainly focussing on the overflow dam and the 5-km-long upstream section of the water-reducing river; no embankments were constructed along the 200-km-long downstream section. Thus the river water spread freely, and the water-reducing river failed to help in reducing the water. The whole water-reducing river project was scrapped due to full sludge during the reign of the Guangxu Emperor.

4.4.2 *The Shaomaying Water-Reducing Dam*

The Shaomaying water-reducing dam is located on the right bank of the South Canal near Zhazi Village, 6 km northwest of Dezhou City. It was built in the 11th year of the reign of the Yongzheng Emperor (AD 1733). In that year, the Weihe River rose and the east bank of the South Canal was broken at multiple locations within the territory of Dezhou. The following year, the Shaomaying Canal overflowed on the east bank. The Governor of Shandong Province, Yue Jun, suggested excavating a new Shaomaying River according to the physical features of the breaches to let the flood reach the Chengong Embankment 10 miles towards the east, entering the old route of the Goupan River from the Cojia breach, continuing into the old Yellow River after flowing to Yuquanzhuang of Wuqiao County in the northeast, merging with the Sinyusi water-reducing river, and finally entering into the ocean. Meanwhile, one stone-rolling dam was planned to be built; it was completed 2 years later. After merging with the Sinyusi water-reducing river, the total 1,798-zhang, 2,994-km-long water-reducing river flowed into the ocean, thereby becoming a total of 160 km long.

In the 13th year of the reign of the Jiaqing Emperor (AD 1808), the river bed was filled with sludge. The Shaomaying water-reducing river (see Figure 4.7) has already been abandoned, and the remains of the watercourse are difficult to find; only a small section of the residual north and south embankments can be discovered in Changzhuang Village. Additionally, stone blocks of the raw materials of the

Figure 4.7: The Ruins of the Shaomaying Water-Reducing Dam on the South Canal (2009)

water-reducing stone dam can be still seen on the embankment of the South Canal within Zhazi Village, north of Dezhou.

4.4.3 *The Jiedi Water-Reducing Gate*

The Jiedi water-reducing gate (see Figure 4.8) is located in Jiedi Town, 7.5 km to the south of Cangzhou City. It was excavated in the 3rd year of the reign of the Hongzhi Emperor in the Ming Dynasty (AD 1490). That year, the Jiedi water-reducing river was excavated and named the 'South Water-reducing River' or 'Brick River'. It flowed northeast into the ocean near present-day Huanghua Qikou from Jiedi Town; it has essentially retained its shape to the present day (see Figure 4.9).

In the 1st year of the reign of the Wanli Emperor (AD 1573), as the embankment of the Zhanghe River was broken and the flood flowed into the Fuyang River, the Jiedi water-reducing gate and others were gradually scraped due to being full of sludge.

In the 4th year of the reign of the Yongzheng Emperor in the Qing Dynasty (AD 1726), Yunxiang, Prince Yi, took charge of excavating Jiedi Pilot and one five-hole stone gate. In the 36th year of the

Figure 4.8: The Jiedi Water-Reducing Gate Room and Gate Hoist (purchased from the German company Siemens in 1933)

reign of the Qianlong Emperor (AD 1771), the gate was changed into the rolling dam. In the later stages of the Qing Dynasty, the water-reducing river was dredged multiple times, continuously embanked, widened, and thickened.

In the 22nd year of the Republic of China (AD 1934), the flood-discharge gate of the Jiedi water-reducing river was changed from five holes to eight holes, enabling the flood-diversion flow to reach 180 cubic metres per second. Meanwhile, a two-hole intake gate was built in the south of Cangxian Village, on the right bank of the Jiedi water-reducing river.

According to the records of the Zhanghe River, the Weihe River and the South Canal Annals, the remains of the works completed in the Qing Dynasty can be still found along the Jiedi water-reducing river. Some bricks in the gate are engraved with characters: 'Made by Wu Qicheng with the leadership of Zhang Ze in Linqing brick kiln in the Yongzheng tenth year'.

Figure 4.9: The Jiedi Water-Reducing River Under the Water-Reducing Gate (2007)

4.4.4 *The Xingji Water-Reducing Gate*

The Xingji water-reducing gate is located between Cangzhou and Qingxian County and was excavated in the 3rd year of the reign of Chenghua in the Ming Dynasty (AD 1467). That year, when the Jiedi water-reducing river was excavated, the Xingji water-reducing river was excavated approximately 2.5 km away from Xingji County; one stone gate was built in the river mouth, 25 km away from the Jiedi water-reducing river in the south, and the two parallel rivers merged downstream. The work was completed hastily and carelessly, so the two water-reducing rivers failed to exert their functions effectively. In the late Ming Dynasty, the two rivers were gradually silted up.

In the 4th year of the reign of the Yongzheng Emperor in the Qing Dynasty (AD 1726), Yunxiang took charge of building one Xingji Stone Dam again and excavating 45 km of the Pilot River to the ocean near Qikou. In the 20th year of the reign of the Qianlong Emperor (AD 1719), the flashboards were added for the Jiedi and Xingji Stone Dams and opened or closed according to the specific situation. To reinforce management on two water-reducing rivers, two Canal Directors were appointed to every 5 km of the canal.

In the 36th year of the reign of the Qianlong Emperor in the Qing Dynasty (AD 1771), in order to reduce inundation in Tianjin, the Qianlong Emperor commanded the changing of the Xingji Gate into an overflow dam and the reduction of the elevation of the top of the dam by 1 chi (1 chi roughly equals 0.33 m). In around the 12th year of the reign of the Jiaqing Emperor (AD 1807), the river was silted up and the dam was lower, the dam top being heightened by 2 chi (0.66 m). In the late Qing Dynasty, the Xingji water-reducing river was gradually abandoned due to sludge.

After a flood in 1963, during the process of launching water-conservancy projects to fundamentally solve the problems in the Haihe, the Xingji water-reducing river was replaced by the new Ziya River, which generally passed along the south bank of the Xingji water-reducing river. As the method of cutting bends and widening had been executed before during the expansion of the South Canal, the original dam site of the Xingji water-reducing river was searched to find a similar solution. Jiang Chao undertook a field visit in the 1980s and judged that the Xingji Dam should be located in the south of modern-day Qingxian County, but in the west of Fengguantun Village, near the Luoguo Bridge on the left bank of the South Canal. Today, one section of the abandoned watercourse of the Xingji water-reducing river can still be seen in the north of Shangwu Village, Qingxian County, with the embankment and river bed still obvious.

4.4.5 *The Machang Water-Reducing Gate (Jiuxuan Gate)*

The Machang water-reducing gate, the last flood-diversion appliance on the South Canal to be completed, was built in the 6th year of the reign of the Guangxu Emperor in the Qing Dynasty (AD 1880).

In the early years of the Guangxu Emperor's reign, the Sinyusi, Shaomaying, and Xingji water-reducing rivers on the South Canal were abandoned due to sludge; only one river in Jiedi was available, and it stretched for only 20 km. Approximately 35 km of the downstream section of the canal was almost silted up. To solve problems such as frequent embankment breakage of the Weihe River and the like, in the 6th year of the reign of the Guangxu Emperor (AD 1880), the Zhili Governor Li Hongzhang deployed around 30 battalions of soldiers (one battalion contained 850–900 soldiers) of Zhou Shengchuan's Huai Army in Machang to excavate the Machang water-reducing river. The 30-km-long water-reducing river flowed to Xinnong Town through the Beitangwafu Civil Gate in Tianjin County. It was connected with the new 45-km-long straight river, which was excavated by Zhou Shengchuan's Army, and finally entered the ocean, thereby lengthening to 90 km as a compound channel. This not only increased the amount of water it could carry, but also prevented the risk of collapse of the main embankment if it became water-logged. A large stone five-hole bridge gate was built at the entrance of the water-reducing river. At that time, no matter if the canal floodwater rose or fell, no less than five flashboards were allowed. As a result, three flashboards were put down throughout the year, coming to a height of 4.5 chi (1.5 m). The bottom of the water-reducing river was 4 chi (1.3 m) higher than bottom of the canal, making the depth of the water in the South Canal more than 8 chi (2.7 m).

In the late Qing Dynasty, after canal transport was stopped, the river was gradually silted up and became dry, leaving it at high risk of inundation. A flood occurred in the 6th year of the Republic of China (1917); the following year, the Shunzhi Water Conservancy Committee changed the old rolling dam in the upstream river mouth of the Machang water-reducing river into a new gate. Two years later, a new, 13-km-long Machang water-reducing river was excavated and located on the upstream south bank of a small station of the Machang water-reducing river. A six-hole check gate was built where the flow reached a rate of 100 cubic metres per second

Figure 4.10: The Jiuxuan Gate (2010)

to check the amount of floodwater that was being diverted from the South Canal.[3]

The Jiusuan Gate (see Figure 4.10) is now around 100 years old, although the gate and the hoisting equipment have been replaced. Other complete buildings have not been reconstructed, but they still perform their functions.

4.5 Summary

The South Canal enables all the branch streams of the southern system of the Haihe River to be connected as the trunk stream of

[3] *Tianjin Water Resource Annals*, compiled by the Water Resource Annals Compiling Committee of the Tianjin Water Resource Bureau and published by Tianjin Science and Technology Press in Tianjin: 2003: 636.

the southern system of the Haihe River, and the Canal still plays a role in preventing floods, draining flooded fields, and so on even today.

During the Yuan, Ming, and Qing dynasties, many solutions were enacted to solve the South Canal's problems. For example, when there was a greater waterway longitudinal slope and a large change in the water levels in the flood and dry seasons, the longitudinal slope of the waterway was effectively decreased. The flow rate was lowered, the amount of water in the transport channel was controlled, and the quality of navigation along the canal was improved through measures such as the alteration of the remaining bends. The intentional alteration of these bends reflects the scientific nature of canal work planning in the ancient times.

The flood-diversion project was the first built within the section of the South Canal and the North Canal of the present-day Yellow River. To control the amount of water in the canal and meet the comprehensive demands of damming on the low water level and flood diversion during the flood period, the Sinyusi, Shaomaying, Jiedi, Xingji, and Machang check hubs and others were arranged in sequence along the South Canal from south to north. The floodwaters were divided by the water-reducing dams, discharging the excess water east into the ocean through the water-reducing rivers. This type of work was integrated with hubs, dams, and gates, reflecting the scientific planning, design, and structure of hydraulic apparatus in the ancient times.

To solve the problems on the South Canal of flood resistance pressure and frequent emergencies during the flood seasons, control and guidance works were arranged along the dangerous work sections. During river-channel project management, the measures of carrying away the sludge through the water of the Zhanghe River (which was high in sand content), reinforcing the embankments with sand, and so on were always adopted; these operations reflect the characteristics of river regulation through the embankments and through other rivers by engineers in ancient times.

Table 4.1: The South Canal's Legacy

River section	Category		Description
South Canal	River	Trunk stream	South Canal
		Water-reducing river	Sinvsi water-reducing river
			Jiedi water-reducing river
			Machang water-reducing river
	Controlled work	Gate dam	Sinvsi water-reducing dam
			Shaomaying water-reducing dam
			Jiedi Gate
			Machang Junction (Jiuxuan Gate)

The river configuration and the natural embankments of the South Canal's bends have been maintained up to now, forming unique bank landscapes. The works on the bends and the dam type water-reducing gates are the enduring legacies of the ancient South Canal (see Table 4.1).

Chapter 5

The Huitong Canal (Including the Middle Canal)

5.1 Introduction

The Huitong Canal is also known as the Shandong Canal. It is the most important section of the Grand Canal, connecting the Haihe River, the Yellow River, and the Huaihe River. In 1128, the Yellow River burst its banks at Kaifeng, Henan Province, and changed its course; it would flow south for the next 700 years. After the Yellow River changed its course, the reach of the canal between the Yellow River and the Huaihe River was occupied by the new channel of the Yellow River. As a result, the waterway connecting the north and the south was interrupted. After the capital of the Yuan Dynasty was established in Beijing in 1172, it became a matter of urgency to restore the waterway system connecting the north and south so as to ensure the food supply for the capital city of Beijing. As a result, the construction of the Grand Canal — with Beijing as its terminus of transportation — began. In the Yuan Dynasty, two new canals were constructed and connected to the previous canal, which, as a whole, formed a new Grand Canal that passed through the northern and southern areas in the east of China (see Figure 5.1). The two new canals were the Tonghui Canal and the Huitong Canal, which were built in the 13th century.

The Huitong Canal built in the Yuan Dynasty started at Linqing, Shandong Province, where it was connected with the South Canal (the southern part of the Grand Canal), and ended at Xuzhou, Jiangsu Province, where it was connected with the Yellow River. The Yellow River was connected with the Huaiyang Canal at Huai'an, Jiangsu Province. Therefore, from the 13th to the 16th century, the waterway between Xuzhou and Huai'an was the new channel of the Yellow River, which was a reach of the Grand Canal. The Yellow River

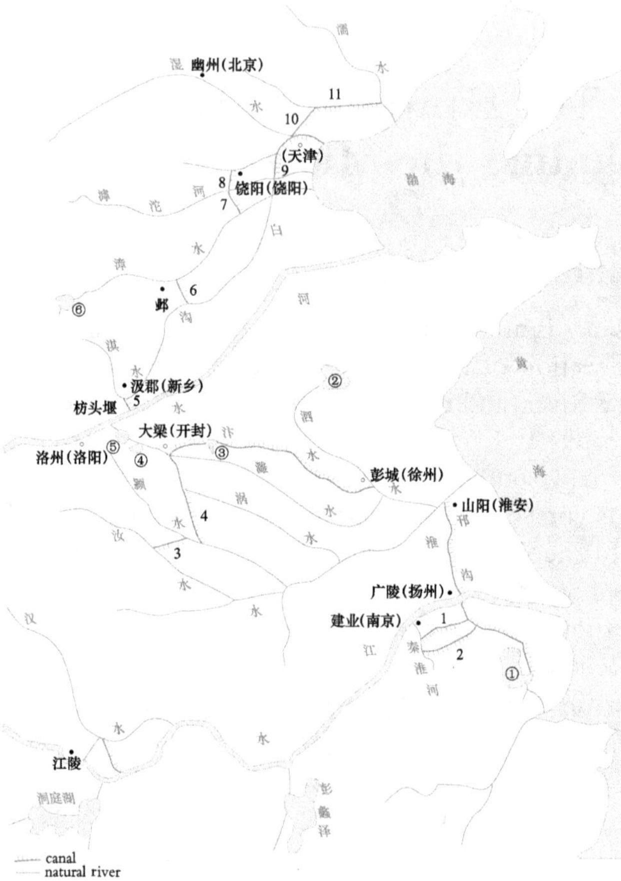

(a) Huitong Canal and Middle Canal (between the 19th century and the 50s of the 20th century)

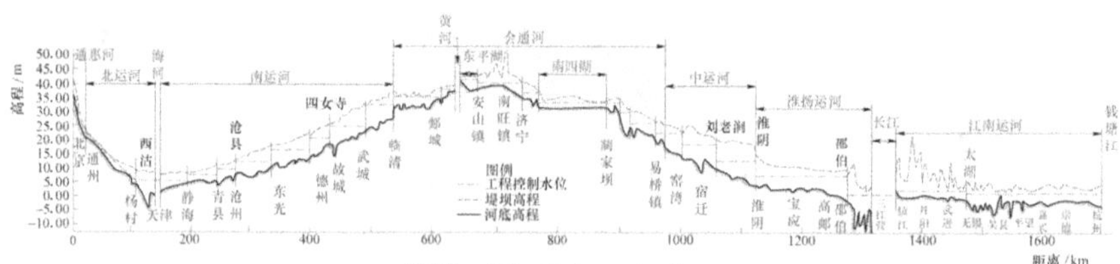

(b) Places Huitong Canal passes through

Figure 5.1: The Huitong Canal

had never been peaceful after it changed course to flow south. A huge amount of sediment in the Yellow River caused serious sedimentation problems within less than 200 years, resulting in levee breaches and flooding when large floods occurred. The Huitong Canal was the channel that was affected by the Yellow River the most. The riverbed aggradation and levee breaches of the Yellow River greatly disturbed the operation of the Grand Canal. After the 16th century, the southern part of the Huitong Canal — the reach affected by the Yellow River — gradually changed its course, which finally led to construction of a new canal, namely the Middle Canal. The Middle Canal was connected with the Huitong Canal at the Weishan Lake, passed through Suqian of Jiangsu Province in the east, and was connected to the Yellow River in Huai'an.

During the 700-year-period in which the Yellow River flowed south, frequent levee breaches and flooding created a profound impact on the adjacent Huaihe River basin, including the change of the Huitong Canal channel, the construction of the Middle Canal, and enormous changes of the river and lake systems connected with the canals. Over several hundred years, a number of lakes formed between the Huitong Canal and the Middle Canal, and some tributaries of the Huaihe River evolved into an independent water system. In 1855, the Yellow River burst its banks at Tongwaxiang, Lankao County, in Henan Province, restoring the waterway into Bohai Sea in the north. Afterwards, the reach of the Huitong Canal in the north of Jining was abandoned, whereas its reach in the south of Jining remained China's most important inland waterway (see Figure 5.1).

5.2 History

In 1128, the Yellow River burst its banks at Kaifeng, Henan Province, and changed its course, beginning a period of 700 years during which the river flowed south. After the Yellow River changed its course, its new channel passed through Bianshui and Sishui, joined the Huaihe River at Huai'an, and finally flowed into the sea via the Huaihe River. The original canal connecting the Yellow River and the Huaihe River, the Tongji Canal (or the Bian River), Sishui (the tributary of the Huaihe River), and the lower reach of the Huaihe River became the channels of the Yellow River.

After the capital of the Yuan Dynasty was established in Beijing in 1172, it became a matter of urgency to restore the waterway system connecting the north and south so as to ensure the food supply for the capital city of Beijing. After the Yellow River changed its course, the original canal system was damaged and two new canals connecting the Jiangnan region (south of the Yangtze River) and Beijing started to be constructed. One of the new canals was the Tonghui Canal, and the other was the Huitong Canal. The Huitong Canal passed the uplands of Southwest Shandong, connecting three large water systems — the Yellow River, the Huaihe River, and the Haihe River. In 1275, Guo Shoujing, a scientist in the Yuan Dynasty, designed the canal route. The canal began to be constructed in 1282 and was completed 7 years later. Kublai Khan named the canal the 'Huitong Canal' (see Figure 5.2). The completion of the Huitong Canal led to the formation of the south–north Grand Canal (today's Beijing–Hangzhou Grand Canal) that passed through the Eastern Plain of China.

The Huitong Canal passed through the alluvial fans of the middle of the Yellow River and the western uplands of Shandong Province, with a total length of 350 km. Topographically, the canal had the highest elevation around Nangwang Town, Jining, in Shandong Province, which averaged approximately 43 m. The longitudinal slope of the canal was around 2 per cent. The lowest point in the southern part of the Beijing–Hangzhou Grand Canal was around the canal mouth connected with the Yangtze River in Yangzhou, and the lowest point in its northern part was around Jinghai, Tianjin. The elevation difference between the highest point of the Huitong Canal and the lowest points of the southern and northern parts of the Grand Canal was approximately 50 m. In the Huitong Canal region, precipitation has an uneven seasonal and inter-annual distribution. The average annual precipitation is 836 mm, with most of it concentrated in the period from June to September. The ratio of precipitation in wet seasons to dry seasons is 2:1. Due to the elevation difference and

Figure 5.2: The Ancient Huitong Canal (Between the 13th and 16th Century)

uneven distribution of water resources in the Huitong Canal, facilities for water control and water supply had to be utilised, and projects for water storage had to be constructed to ensure sufficient water for continuous navigation along the canal.

The Huitong Canal is the reach with the most control projects in the Grand Canal. The Huitong Canal crosses the divide between the Yellow River basin and the Huaihe River basin. The water sources along the Huitong Canal are the Yellow River, the Wen River (a tributary of the Yellow River), and the Sishui River (a tributary of the Huaihe River). The Huitong Canal has been used for more than 500 years, mainly thanks to its perfect water supply projects — known as 'water tanks' in Ancient China. Water tanks are equivalent to modern-day reservoirs, and they were built along both banks of the canal between Jining and Nanwang. Water tanks were built by taking advantage of the topography. Weirs were built in the Wen River and the Sishui River, and water from there was diverted to the water tanks on both banks of the canal. Then the water tanks supplied water for canals by means of small sluices. Ships that entered the Huitong Canal were generally organised into a fleet consisting of 200–300 ships and passed through the canal in groups. The gates on the Huitong Canal were operated similar to modern ship locks. The canal between two gates was filled with water from water tanks; by the opening and closing of the gates, ships could pass through the Huitong Canal with a high elevation.

The construction and operation of the Huitong Canal took place in the period during which the Yellow River flowed south (the period between 1192 and 1855, during which the Yellow River flowed into the Yellow Sea through the Bian River, the Sishui River, and the lower reach of the Huaihe River). The flooding and sediment of the Yellow River resulted in a propensity to suffer sedimentation and a change in its course. While it flowed south, the Yellow River frequently burst its banks, changed its course, or bifurcated into several branches, which resulted in large floods and a large amount of sediment. Therefore, the canals were often interrupted or silted up, and the natural conditions of the area that the Huitong Canal passed through were gradually changed. During the 700 years from the 12th to the 19th century, the Yellow River levee was built to control floods in the Yellow River, blocking the lower reach of the Sishui River, a tributary of the Huaihe River. As a result, the Dushan Lake, the Zhaoyang Lake, the Nanyang Lake and the Weishan Lake (generally referred to as the 'Southern

Four Lakes') were formed. The Sishui River no longer flowed into the Huaihe River but changed its course, flowing southeast along Yimeng Mountain, joining two other tributaries of the Huaihe River — the Yi River and the Shu River — and finally flowing into the sea. The Yi River, the Shu River, and the Sishui River have been called the 'Yi–Shu–Si Water System'. In the middle of the dramatic change in the northern water system of the Huaihe River, the Huitong Canal changed as well. From the middle of the 16th century, the canal mouth where the Huitong Canal was connected with the Yellow River moved eastward constantly. It was not until the Middle Canal was excavated that the Huitong Canal was connected with the Middle Canal via the Weishan Lake.

The Middle Canal is the last reach that was formed in the entire Beijing–Hangzhou Grand Canal. In the middle of the 16th century, the Yellow River changed greatly. Xuzhou-Hong and Lyuliang-Hong, two places with rapids and dangerous shoals, were changed into shallow reaches. The waterway for tribute rice transportation in the Yellow River was blocked frequently because the river dried up. In July of the 3rd year of the reign of the Longqing Emperor in the Ming dynasty (AD 1568), the river burst its banks at Peixian County and, thus, the reach from the mouth of the Huitong Canal at Xuzhou to Tea city was silted up. In the same year, more than 2,000 ships for the tribute rice transportation were blocked in Pizhou City. Weng Dali, the then-top official in charge of rivers, put forward the planning of the excavation of the Jia River — the excavation of a new canal in the estuary of the Xue River that was connected with the Jia River via the southeast of Hanzhuang and flowed south into the Yellow River. Because the project was very big, its construction was not started until the 28th year of the reign of the Wanli Emperor (AD 1600), 30 years later, and was completed in the 32nd year of his reign (AD 1604). Initially, the Jia River flowed directly into the Yellow River at Pizhou. In the 5th year of the reign of the Tianqi Emperor (AD 1625), a new channel was excavated that flowed southeast to Suqian, passed through the Huangdun Lake and the Luoma Lake, and finally flowed into the Yellow River. As a result, the canal mouth was connected with the Yellow River through multiple channels.

In the 25th year of the reign of Emperor Kangxi (AD 1686), Jin Fu, a top official in charge of rivers, led the work of extending the Jia River eastward to excavate the Middle River that was later called the Middle Canal. The Middle Canal was excavated in the low-lying land between the Lu Levee and the Yao Levee in the north bank of the Yellow River and was extended eastward until it reached the east bank of Qingkou, Huaiying. After the completion of the Middle River, the Grand Canal only intersected with the Yellow River at Qingkou; except for that, the Grand Canal was separated from it. Afterwards, ships for the transportation of tribute rice crossed the Yellow River at Qingkou and, upon leaving Qingkou, entered the Middle Canal via the Zhongjiazhuang Sluice, which avoided the risk of sailing 100 km in the Yellow River and being affected by the winds and waves in it. The time that it took for ships to travel from the Huaihe River to Tongzhou was a month shorter than before the Middle Canal was constructed.

The water sources for the Middle Canal were from the Sishui River, the Yi River, and the Shu River. Originally, the Sishui River, Yi River, and Shu River were the tributaries of the Huaihe River. After the completion of the Middle Canal, however, they became an independent water system — namely the Middle Canal Water System. At present, it is also called Yi–Shu–Sishui Water System.

The Yi–Shu–Sishui Water System has its source in Yimeng Mountain. Some streams in Yimeng Mountain converged, and thus a seasonal lake was formed. By constructing levees and control sluices around it, the lake became an artificial lake with the function of regulating water, namely the Luoma Lake. The Luoma Lake was not only a reliable water source for the Middle Canal, but also a reach of the Middle Canal. The Middle Canal passed through Taierzhuang of Weishan County in Shandong Province, flowed east, then passed respectively through Pizhou, Suqian, and Huai'an in Jiangsu Province, and continued to flow to Yangzhuang, where it bifurcated into two branches; one was the Liutang River and the other was the main line of the Middle Canal. The Liutang River was a channel connecting the Yi–Shu–Sishui water system and the sea, and the main function of the Liutang River was to discharge floods.

The main line of the Middle Canal joined the Yellow River at Yangzhuang, from which, across the Yellow River, was the canal mouth where the Huaiyang Canal intersected with the Yellow River. The Middle Canal had a navigational length of 180 km and a width of 50 m. The reach in the Luoma Lake had a maximum width of 1,910 m. Its longitudinal slope was 0.1 per cent.

In 1855, the Yellow River burst its banks at Tongwaxiang, Henan Province. Since then, it had returned to its north course. After the Yellow River burst, the Huitong Canal was interrupted at Zhangqiu, Shandong Province, and the Yellow River flowed through the Daqing River and into the sea at Lijin, Shandong Province. Soon after that, the system of tribute rice transportation that had lasted for more than 2,000 years was ended and the mission of the Huitong Canal, a project whose maintenance was extremely difficult, was accomplished. The Huitong Canal then gradually evolved into a regional river. At present, to the south of Wenshang in Shandong Province, the Huitong Canal belongs to the Yi–Shu–Sishui river basin, which belonged to the Huaihe River basin before Yellow River diverted southwards into the Huaihe River. Meanwhile, to the north of Dongping, it belongs to the Yellow River basin.

The construction and operation of the Huitong Canal and the Middle Canal were closely related to 700 years of evolution of the Yellow River. The two canals witnessed the impact of the Yellow River on the Huaihe River while the Yellow River flowed south. They were also the products of the environmental evolution of rivers. As a result, they have the dual characteristics of cultural and natural heritage.

5.3 Pivotal Control Projects on the Huitong Canal

The Huitong Canal runs through the hilly area in the southwest of Shandong Province. Its north end is connected with the water system of the Haihe River and its south end with the water systems of the Yellow River and the Huaihe River. The water sources of the Huitong Canal were from the Wen River of the tributary of the Yellow River, the Yellow River, and the Sishui (today's Si River), Yi, and Shu Rivers of the tributaries of the Huaihe River. The key projects of the Huitong Canal in a north-to-south order were: the Linqing Hub Project, which was located at the north end of the Huitong Canal and connected with the Wei River and the South Canal; the Daicun–Nanwang Hub Project, which was the water-source project in the reach passing through the divide; the Jining Hub Project, which intersected with the Sishui River; the Xuzhou Hub Project intersecting with the Yellow River; and the continuous control sluices in the canal. The Xuzhou Hub Project was at the south mouth of the Huitong Canal and intersected with the Yellow River, where the Huitong Canal and the Yellow River had almost the same water level. The door of the Xuzhou Hub Project consisted of weir and sluices that were operated for shipping. As the southern reach of the Huitong Canal changed its course at the end of the 16th century, the south mouth of the canal was gradually abandoned and sealed shut by the construction of a levee.

5.3.1 *The Linqing Hub Project at the Intersections of the Huitong Canal, the Wei River, and the South Canal*

The Huitong Canal and the Wei River converge at Linqing. The Wei River was both the water source of the Huitong Canal and a river with the function of navigation. At the confluence of the Wei River and the Huitong Canal, a navigational hub project was constructed as an important joint of the Huitong Canal (see Figure 5.3). At Linqing, ships could enter the Wei River and then travel to Daming (modern-day Daming and Shijiazhuang, Hebei Province) and Weihui (modern-day Xinxiang and Jiaozuo, Henan Province). Also at Linqing, ships could enter the South Canal directly and travel to Beijing via Cangzhou and Tianjin. The Linqing Hub Project was constructed in the Yuan Dynasty, constantly improved in the Ming Dynasty, and then used until the end of the Qing Dynasty. At Linqing, Huitong Town was established in the Yuan Dynasty and Linqing County was established during the reign of the Jingtai Emperor in the Ming Dynasty, with Huitong Town as the

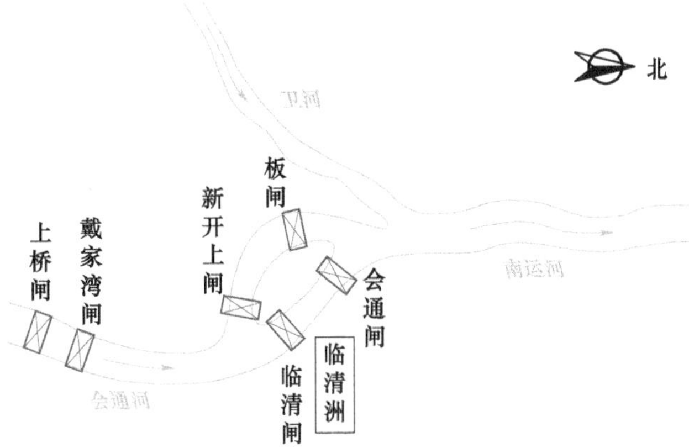

Figure 5.3: The Linqing Hub Project Connecting the Huitong Canal, the Nan Canal, and the Wei River

capital of the county. During the Hongzhi Emperor's reign in the Ming Dynasty, Linqing County was promoted to Linqing Prefecture, with Huitong Town as the capital of the prefecture. In the Jiajing Era of the Ming Dynasty, city walls were constructed around the Linqing Hub Project, and the Linqing Hub Project became a shipping port for the transportation of tribute rice. As a result, Linqing became an important city with a prosperous business in the North China Plain.

At the confluence of the Huitong Canal, the Wei River, and the South Canal, it was difficult for ships to pass through because of the difference between the elevations of the three rivers. The water source of the Huitong Canal was the Wenshui River, which offered a small amount of water, while the Wei River and the South Canal were the main rivers in the southern part of the Haihe River, and they were rich in water resources. Because of the large difference between the hydrological characteristics of the Huitong Canal and the South Canal, the mouths of the two canals could not make a seamless whole in terms of water level. The Linqing Hub Project consisted of a water-dividing weir, a branch channel, and four sluices (see Figure 5.3). Through these engineering facilities, the conditions of water flow and navigation were greatly improved. The Huitong Canal was bifurcated into the north and south branches by the water-dividing weir (named 'Aotouji'). The south branch joined the Wei River and the north branch joined the South Canal. There were the Huitong Sluice and the Linqing Sluice on the north branch, while the Xinkai Sluice (also known as the 'Brick Sluice') and the Ban Sluice were on the south branch. Two pairs of sluices were operated together to control the water depth and volume of the Huitong Canal, allowing the ships transporting tribute rice to pass through the canal mouth smoothly.

The Huitong Canal was completely formed in 1289. Four years after the formation of the Huitong Canal, the Huitong Sluice was constructed. Three years after the construction of the Huitong Sluice, the Linqing Sluice was constructed. At the time, the Linqing Hub Project was partially formed. In 1417, Chen Xuan led the excavation of a branch channel (called the 'Yue River') south of the Aotaoji Weir to connect the Huitong Canal and the Wei River and organised the construction of the Ban Sluice on the Yue River, 3.5 km southwest of Linqing City. Through such an improvement, the longitudinal slope of the Grand Canal near the Linqing intersection was reduced and the navigational conditions at the Linqing Canal Mouth were improved. In 1432, the Ban Sluice was converted into a masonry sluice. During the period of 1457–1464, the Linqing Sluice was abandoned. In 1490, Bai Ang rebuilt the abandoned Linqing Sluice and moved it southward for approximately 300 m.

The Ban Sluice was appropriately located so it could protect the mouth of the Huitong Canal effectively. When the Wei River had a high flow, the Ban Sluice was closed to avoid the impact of the high flow on the Huitong Canal. When the Huitong Canal had a low flow, the Ban Sluice was opened to divert water from the Wei River to the Huitong Canal. The Wei River was a sediment-laden river, so the reach around the canal mouth was often silted up. The Ban Sluice could largely reduce the impact of sedimentation on the canal mouth. With the Huitong Sluice and the Linqing Sluice as one pair and the Xinkai Sluice and the Ban Sluice as the other pair, two sets of conjunctive operations were formed, which not only balanced the difference between the water levels of the upstream and downstream sluices, but was also beneficial for the river dredging around the Linqing Hub

Project and the maintenance of the sluices, thus prolonging the operational time of the sluices to the largest extent possible.

At present, the *status quo* ante of the Linqing Hub Project remains unchanged. There is no significant change in the canal routes. The Aotouji Weir and the two branches on both sides of the Aotouji Weir still exist. So far, the Linqing Sluice, the Huitong Sluice, and the Xinkai Sluice have been kept intact (see Figure 5.4).

5.3.2 *The Daicun–Nanwang Hub Project*

The Daicun–Nanwang Hub Project diverted water from the Wen River to the Huitong Canal. It consisted mainly of Daicun Weir, a diversion channel (also called 'Small Wen River'), Nanwang Water Tanks, and two control sluices (see Figure 5.5). Around Nanwang Town in Jining, Shandong Province, the Huitong Canal had the highest elevation in the entire Grand Canal. The Wen River, with mountainous streams as its water sources, flowed through Ningyang and finally into the sea in the east of Dongping (in 1855, the Yellow River changed its course and flowed into the sea via the Daqing River, so the Wen River became a tributary of the Yellow River).

The Daicun Weir was built across the Wen River and, thus, the water level of the Small Wen River was heightened. Through the Small Wen River, water was diverted from the Wen River to the Nanwang Water Tanks — Shushan Lake, Nanwang Lake, and Mata Lake. The water in the Nanwang Water Tanks flowed into the Huitong Canal through small sluices and then was diverted to the north and the south respectively by Liulin Sluice and Shili Sluice (Figure 5.5). The Daicun–Nanwang Hub Project ensured water supply for the reach of the Huitong Canal passing through the divide.

The Shushan Lake, the Nanwang Lake, and the Mata Lake at Nanwang are ancient lakes resulting from the levee breaches of the Yellow River in the 10th century, and they were collectively known as the Liangshan Lakes. The builders of the Huitong Canal used the low-lying lakes as the water tanks to supply water for the Huitong Canal. The three lakes had a total water surface area of 188.5 km^2.

(a) Linqing Hub Project

(b) Status quo of the brick sluice (xinkai sluice in the Ming and Qing Dynasties) (Taken in 2008)

Figure 5.4 The Linqing Hub Project and its Control Sluices

Note: The painting shows the Linqing Weir built of bricks and Huitong Town in the Qing Dynasty, and also shows the conjunctive operation of the Huitong Sluice and the Linqing Sluice and as the relationship between the Huitong Hub Project and the Huitong Canal/Wei River. It also reflects the social flavour along the canal near Linqing City in the Qing Dynasty.

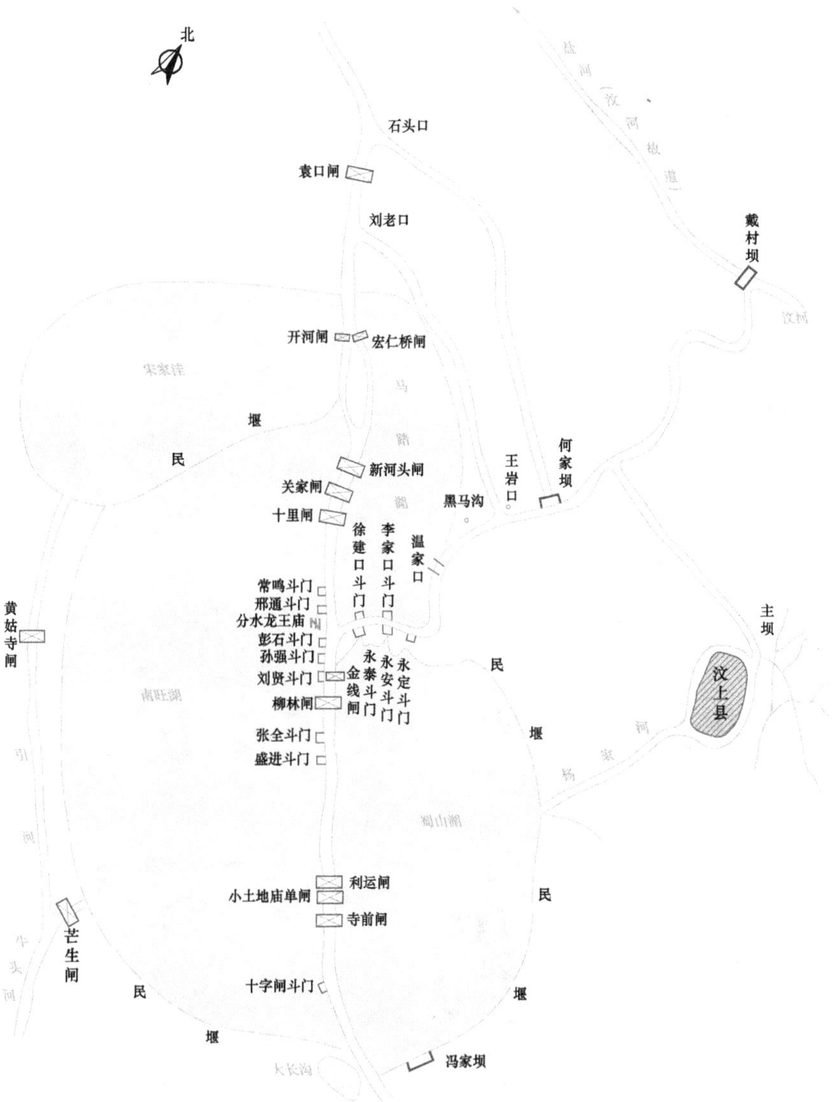

Figure 5.5: The Daicun–Nanwang Hub Project (From the 15th Century to the 19th Century)

Given an average water depth of 2 m, their total storage volume was 3.7×10^8 m^3. Water was diverted from the Wen River to the water tanks. By regulating water in the water tanks, the Huitong Canal could be kept navigable for nine to 10 months each year.

5.3.2.1 *Daicun Dam*

Daicun Dam (see Figure 5.6) was built in 1411. It was originally an earth dam. Approximately 180 years later, it became a barrage consisting of a rockfill dam and a masonry dam. Through the construction of the Daicun Dam and the excavation of a diversion channel, water was

(a) Current Daicun Dam

(b) Foundation of Daicun Dam

Figure 5.6: Daicun Dam

diverted from the Wen River to Nanwang so as to supply water for the Huitong Canal. Such a perfect engineering system originally came from the idea of an old man named Bai Ying, who was in charge of river management in Wenshang County. Later, the diversion channel from Daicun Dam to Nanwang was called the Small Wen River. After its completion, Daicun Earth Dam was broken by floods every year and reconstructed the following year. The Wen River was short, with a rapid flow. It took a lot of manpower and material resources to build the dam on the Wen River every year. As a result, in 1572, Daicun Earth Dam was converted into a rockfill dam with a width of 500 m, a length of 700 m, and a height of 5 m. The rockfill dam was actually a dam that could both retain water and allow water to permeate through it. In 1588, Daicun Rockfill Dam was converted into an earth dam with a masonry spillway. At that time, Daicun Dam had a length of 370 m, a crest width of 5 m, a bottom width of 6 m, and a height of 5 m, and the length of the spillway was 200 m. The side of the dam facing water was built of boulder and the dam face was a long and gentle slope. The dam body was built of boulders with a grain size of approximately 3 m, and the inside of the dam was filled with fine stones.

After the 17th century, Daicun Dam was converted many times. Eventually, the earth dam in the middle part of Daicun Dam was demolished. Instead, a permeable rockfill dam was built in the middle, with a masonry dam at either end of it. At that time, Daicun Dam had a total length of 350 m and a height of 1.7 m. The permeable dam in the middle of the river had the function of breaking automatically, which reduced the risk of the whole dam being broken by floods. If the dam was damaged partially, it could be repaired quickly. The unique dam type enhanced its benefits effectively.

5.3.2.2 *Nanwang Water Tanks and Their Use*

Daicun Dam diverted water from the Wen River to the Nanwang Tanks via the Small Wen River. Originally, the Nanwang Water Tanks used low-lying lands to store water with no engineering facilities, and therefore their regulation capacity was low. When there were high flows in the Wen River, the Nanwang Tanks became a large area of water, while when there were low flows in the Wen River, the Nanwang Water Tanks became small ponds, which resulted in the Huitong Canal drying up. In the middle of the 15th century, Nanwang South Sluice and Nanwang North Sluice were constructed on the Nanwang Channel and levees around the lakes were constructed as well, which allowed the Nanwang Water Tanks to have an annual regulating capacity. As a result, the project that diverted water from the Wen River to the Huitong Canal could fulfil its full potential and create an economic benefit.

The Nanwang North Sluice was later called 'Shili Sluice', while the Nanwang South Sluice was called 'Liulin Sluice'. The completion of the two sluices first determined how much water in the Huitong Canal was diverted to the north by Shili Sluice and how much water was diverted to the south by Liulin Sluice. Moreover, additional continuous sluices were constructed upstream of Liulin Sluice that could control the water from the upstream reaches and keep the necessary water depth for navigation between Liulin Sluice and Shili Sluice.

The Wen River was a mountain stream in which the flood volume in flood season tended to account for 70 per cent of that of the whole year. As a result, it was necessary to construct dams on the Wen River to store water in it and increase the storage capacity of the Nanwang Water Tanks. Topographically, the Huitong Canal was higher than the Nanwang Lake (the Nanwang Water Tanks), so water in the Nanwang Lake could not flow into the Huitong Canal. As a result, it was necessary to construct sluices to raise the water level in the Nanwang Lake by storing water in it. Since the construction of the project to divert water from the Wen River to the Huitong Canal in the 9th year of the Yongle Era (AD 1411), many water-control projects had been built in the Nanwang Lake. Approximately 100 years later, the Nanwang Lake had a perimeter of 25 km. A channel crossed the lake with two long levees along it. There were small sluices and bridges on the west levee. The bridges were used for boat trackers. To the west of the small sluices were water tanks for storing water. In the 4th year of the Chenghua Era (AD 1468), Chen Shan, an official in Shandong, led the construction of the west levee by using stones, because the original west levee was made of earth and prone to damage. The east levee was also rebuilt. After 95 years of construction (1411–1505), the Nanwang Water Tanks became a water source project with the basic functions of modern reservoirs. Levees were built around the Nanwang Lake and there were sluices on the levees to control and regulate water in the lake.

From the Jiajing Era to the Wanli Era in the Ming Dynasty, the Nanwang Project experienced an essential repairs three times. In the 13th year of the reign of Jiajing (AD 1534), Liu Tianhe repaired approximately 25 km of levee around the west part of the Nanwang Lake and 18 sluices. In the 20th year of the reign of the Jiajing Emperor (AD 1541), Wang Yizhan, the Vice Defence Minister, led the construction of channels and small sluices along the levee of the Huitong Canal and the excavation of more than 20 ditches crisscrossing the Nanwang Lake. The small sluices were used to divert water from the Nanwang Lake to the Huitong Canal. Around the Nanwang Lake, willows were planted and stelas were set up to prohibit people from reclaiming the lake. In the sixteenth year of the reign of the Wanli Emperor (AD 1588), Chang Jujin and Pan Jixun, two high-ranking officials, increased the levee of the Nanwang Lake by approximately 47 km and built a levee with a length of 7 km, a bottom width of 5 m, a crest width of 2.7 m, and a height of 2.7 m in the Nanwang Lake. At the time, there were 14 small sluices connecting the Nanwang Lake and the Huitong Canal. The improved water control facilities in the Nanwang Lake in the Ming Dynasty were still used in the Qing Dynasty.

The Wen River was a sediment-laden river. One hundred years after the construction of Daicun Dam, during the Jiajing Era (1522–1566), the Nanwang Water Tanks were seriously silted up. Liu Tianhe, an expert from the Ming Dynasty, stated:

> In Chenghua's Period, the reach downstream of Daicun Dam was not seriously silted up. In Jiajing's Period, however, the reach downstream of Daicun Dam was silted up so seriously that sediment was transported to Nanwang. The Wen River was full of sediment, so it overflowed its banks easily and floods occurred frequently. Sediment deposited on both banks of the Wen River. Winds made the sediment fly in the sky and finally fall back into the river.

During the Qing Dynasty, the reach of the Huitong Canal at Nanwang was so full of sediment that it was extremely difficult to dredge the canal. Every time dredging was carried out, tens of thousands of people were employed and a lot of money was spent.

By constructing levees and sluices at Nanwang, water diverted from the Wen River was first stored in the Nanwang Water Tanks and then provided to the Huitong Canal, which greatly reduced the sediment from the Wen River to the Huitong Canal. Therefore, the Nanwang Water Tanks as water sources also had the benefit of reducing sediment. In the flood season of the Wen River, the sluice gates of the Shushan Lake and the Mata Lake were opened to store water diverted from the Wen River. After sediment was deposited in the lakes, clear water was diverted to the Nanwang Lake and finally to the Huitong Canal. The two lakes played an important role in not only storing water but also depositing sediment, so they were also called 'sand tanks'. Because of such sand tanks, dredging was conducted intensively for the sand tanks rather than for river channel. Due to a huge amount of sediment at Nanwang, it took a great deal of effort to dredge it, but since the dredging work was concentrated in the lakes, the difficulties in dredging were largely reduced.

5.3.3 *The Jining Hub Project*

In the 9th year of the Yongle Era in the Ming Dynasty (before the construction of Daicun Dam and the Nanwang Hub Project), the Jining Hub Project was important for the water supply and diversion of the Huitong Canal. The Wen River, the Sishui River, and the Yellow River joined the canal at Jining. Then, by the regulation of three sluices — the upper sluice, the middle sluice, and the lower sluice — water was diverted to the south and the north, respectively.

The Wen River flowed from Gangcheng Dam at Ningyang to Jining via the Guang River, while the Sishui River flowed from Jinkou Dam at Yanzhou to Jining via the Fu River. The Wen River and the Sishui River converged in the south of Jining City, then flowed into Tianjing Sluice (which was called 'Huiyuan Sluice' in the Yuan Dynasty) and finally into the Huitong Canal (see Figure 5.7).

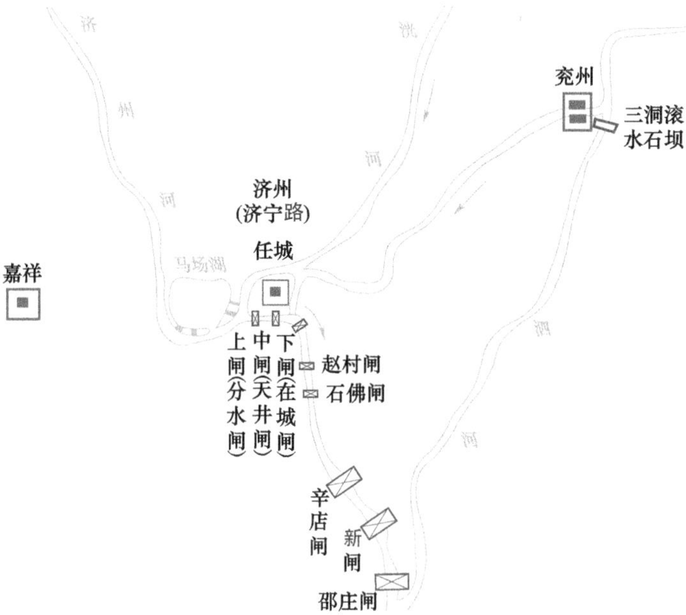

Figure 5.7: Schematic Diagram of the Jining Hub Project, Gangcheng Dam, and Jinkou Dam for Diverting Water from the Wen River and the Sishui River to the Huitong Canal in the 15th Century

According to the records of Jiexisi in the Yuan Dynasty, in the Chenghua Era in the Ming Dynasty, between Tachangkou of Taiyu County and Shuanghekou of Caozhou City, a channel formed by floods of the Yellow River was dredged, Guangyun Sluice was constructed at Tachangkou, and Yongtong Sluice was constructed at Jining. Both of the sluices were on the west bank of the canal and connected with the flood-formed channel. This channel could divert water from the Yellow River to the Huitong Canal.

The Jining Hub Project consisted of the upper sluice (named 'Fenshui Sluice' or 'Diversion Sluice'), the middle sluice (named 'Tianjing Sluice'), and the Lower Sluice (named 'Zaicheng Sluice') (see Figure 5.7). A branch channel was excavated to connect the upper sluice and lower sluice directly. Fenshui Sluice was originally constructed in 1301. Tianjing Sluice (also called 'Yuanhuiyuan Sluice') was 1.5 km away from the upper sluice. After they converged, the Guang River and the Sishui River flowed into Tianjing Sluice from the right bank of the Huitong Canal in the south of Jining City. Zaicheng Sluice was 1 km away from Tianjing Sluice. Before the construction of Daicun Dam and the Nanwang Hub Project, the upper sluice and the lower sluice were the first sluices in the Huitong Canal to divert water to the north and the south respectively. In 1459, three new sluices were constructed in the branch channel connecting the upper sluice and the lower sluice — the new upper sluice near the upper sluice, the new lower sluice near the lower sluice, and the new middle sluice in the middle of the branch channel. Tianjing Sluice was the water intake where the Wen River and the Sishui River converged and flowed into the Huitong Canal through the sluice. Fenshui Sluice and Zaicheng Sluice were used to divert water in the Huitong Canal to the north and the south respectively. The branch channel connecting the upper sluice and the lower sluice became a transit shipping port. Because of the navigational function of the Jining Hub Project, Jining became both a logistics centre in the Ming and Qing dynasties and an important city that boomed due to the canal.

5.3.4 *Projects for Diverting Water from the Wen River and the Sishui River to the Huitong Canal, Including Gangcheng Dam and Jinkou Dam*

Before the construction of Daicun Dam, Gangcheng Dam and Jinkou Dam were the main water source projects on the Huitong Canal. The diversion projects consisted of two parts — Gangcheng Dam, for diverting water from the Wen River to the Huitong River, and Jinkou Dam, for diverting water from the Sishui River to the Huitong River. Water in the Wen River was diverted to the Guang River by Gangcheng Dam at Ningyang and then flowed through the Guang River into Jining. The Sishui River diverted water from the Yi River to Yanzhou. Water in Yanzhou was diverted to the Fu River by Jinkou Dam and then flowed through the Fu River into Jining. After converging in the south of Jining city, the Wen River and the Sishui River flowed into the Huitong Canal via Tianjing Sluice in order to supply water for the Grand Canal.

5.3.4.1 *Gangcheng Dam: A Project to Divert Water from the Wen River to the Guang River*

In AD 1257, Bi Fuguo, a Jizhou official, built a dam in the Wen River in Gangcheng Town, northeast of modern-day Niyang County, and excavated the Guang River to divert water to the Sishui River via Rencheng County (modern-day Jining, Shandong Province). As a result, the Wen River and the Sishui River were connected. To the west of Gangcheng Dam, a water-intake sluice was constructed and later named as the 'West Sluice'. After the excavation of the Huitong Canal, a new sluice was constructed near Gangcheng Dam, which was named 'East Sluice'. Gangcheng Dam was originally a dam built of earth and straw. Every year, it was rebuilt after the flood season. Gangcheng Dam could divert two thirds of the water volume in the Wen River to the Guang River. In 1318, Gangcheng Dam was converted into a masonry dam, but it was damaged soon afterwards by floods. It was later rebuilt by piling up ripraps. In 1338, the sluice for water intake was damaged by floods, and the Wen River changed its course and joined the Guang River, which resulted in serious sedimentation in the Guang River. The following year, a masonry sluice was constructed east of the original site of the sluice.

In 1411, Daicun Dam was constructed 30 km downstream of Gangcheng Dam, but Gangcheng Dam was still left to divert water from the Wen River to the Huitong Canal. In 1473, a masonry dam was constructed 8 km southwest of the old dam site, which was the first permanent masonry dam in the Huitong Canal. The dam had a height of 3.4 m, a bottom width of 7.8 m, a crest width of 5.3 m, and a length of 373 m. There were seven sluices in the dam for water release, each of which had a height of 3.3 m and a width of 3.3 m. The sluice to divert water to the Guang River was 30 m east of the dam. It was a sluice with two sluice gates, each of which had a width of 2. 8 m and a height of 3.42 m.

After the Zhengde Era, Daicun Dam replaced Gangcheng Dam. In the Wanli Era (1573–1617), Gangcheng Sluice was stopped up so that the Wen River no longer flowed south and water was diverted from the Quan River to Jining as an auxiliary water source for the Huitong Canal.

In 1940, Gangcheng Dam was converted into an irrigation project and its related sluices — as well as the Guang River — still functioned normally (see Figure 5.8). Afterwards, its structures on the ground were gradually damaged. In recent years, due to sand excavation in the Wen

(a) Gangcheng Dam

(b) Gangcheng Sluice (in the 1940s)

Figure 5.8: The Gangcheng Sluice and its Water Intake

Figure 5.9: The Ruins of Gangcheng Dam (2008)

River, a part of the dam body, the foundation of the sluice and retaining walls appeared (see Figure 5.9) at the old site of Gangcheng Dam in the Wen River, 14 km northeast of today's Ningyang County. It is known as 'Old Sluice Bend' by the local people.

5.3.4.2 *Jinkou Dam, a Project for Water Diversion from the Sishui River to the Huitong Canal*

The construction of Jinkou Dam began at the beginning of the Sui Dynasty. Xue Zhou, an official in Yanzhou City, built a dam in the Sishui River east of the city to raise the water levels and then excavated a channel to divert water for irrigation. At the time, the channel was called 'Xue's Channel' in Yanzhou. There was a water control sluice at the starting point of the channel. Before the excavation of the Jizhou River in the 21st year of the Yuan Dynasty, Jinkou Dam was an overflow masonry dam. In the early Ming Dynasty, it was an earth dam. In 1472, it was converted into an overflow dam. In the 4th year of the Yanyou Era, in order to construct Jinkou Sluice north of Jinkou Dam, a 30-km-long channel was excavated to divert water from the Sishui River to Huiyuan Sluice in Jining. The channel joined the Wen River at Jining and then flowed into the Huitong Canal (see Figure 5.7). The channel was later called 'Fu River'.

Figure 5.10: Jinkou Dam in the 1930s

In the 1970s, Jinkou Dam (see Figure 5.10) and Jinkou Sluice were reinforced and partially rebuilt. The main function of Jinkou Dam at present is urban and agricultural water supply.

5.3.5 *The South Mouth of the Huitong Canal and the Yellow River*

The reach of the Huitong Canal from Jining to the canal mouth in Xuzhou was formed by improvement of the original Sishui River (see Figure 5.11). From the 13th century to the 15th century, 16 water control sluices were constructed one after another in a 160-km-long reach downstream of Lu Bridge in Jining. The reach became a waterway with a perfect engineering system (see Figure 5.11).

In 1192, the Yellow River changed its course dramatically and the original Sishui River downstream of Xuzhou became the

Figure 5.11: The Reach of the Huitong Canal from Jining to the Southern Canal Mouth from the 13th to the 16th Century

reach of the Yellow River. The reach ran through Xuzhou, Pizhou, Huining, Suqian, and Taoyuan and finally joined the Huaihe River at Huai'an. The Huitong Canal was connected with the Yellow River at the south canal mouth in Xuzhou. The reach of the Yellow River from Xuzhou to Huai'an had a length of 90 km. There were two places with rapids and dangerous shoals in the reach; one was called 'Xuzhou-Hong' (the Chinese character for Hong means 'a place with rapids and dangerous shoals') and the other was 'Lyuliang-Hong'. Xuzhou-Hong was 1 km away from Xuzhou and had a length of 300 m. Lyuliang-Hong was 30 km southeast of Xuzhou and had a length of 2.5 km. There were many large and hard rocks at Xuzhou-Hong and Luliang-Hong. When floods occurred, the rocks were submerged by rising river water, so navigation was not affected. At times of low flow, ships tended to collide with the rocks, which resulted in accidents. From 1403 to 1456, a branch channel was excavated for navigation to avoid ships sailing through Xuzhou-Hong and Luliang-Hong. After the 16th century, the lower Sishui River was silted up due to the sediment from the Yellow River. As a result, navigation in the Sishui River was blocked and lakes were gradually formed south of Jining. The first lakes that appeared were Nanyang Lake, Dushan Lake, and Zhaoyang Lake. Towards the end of the 16th century, Weishan Lake was formed. All the four lakes were collectively called the 'South Four Lakes'. At the time, the Sishui River no longer joined the Yellow River. The South Four Lakes had a water surface area of 1,496.6 km^2 and a total storage capacity of 2.353×10^9 m. During the formation and gradual expansion of the South Four Lakes, the Huitong Canal, which ran through the four lakes, changed its course gradually. At the end of the 16th century, after the Nanyang New River and the Jia River were excavated successively, the mouth of the Huitong Canal in Xuzhou was abandoned. Instead, the Huitong flowed into the Jia River via the sluice in the Weishan Lake and then the Jia River flowed into the Yellow River via the canal mouth at Pizhou (see Figure 5.12).

Figure 5.12: The South Mouth of the Huitong Canal and The Mouth of the Jia River from the 16th Century to the 17th Century

Note: At the end of the 16th century, the course of the lower Huitong Canal was changed. Afterwards, the lower Huitong Canal passed through Hanzhuang Sluice at the outlet of the Weishan Lake, then ran through Taierzhuang, and finally flowed into the Yellow River south of Yaowan Town. In the Jia River, there were multiple canal mouths that connected the Jia River and the Yellow River. After the Middle Canal was excavated in the 18th century, the canal mouths in the Jia River were stopped up. The mouth of the Middle Canal was moved to Yangzhuang in Huai'an.

With its riverbed aggradation, from the middle of the 16th century, the Yellow River lost its navigational function and the Huitong Canal was no longer connected with the Yellow River. In the 1550s, the Jia River was excavated and then improved constantly. As a result, the Grand Canal was gradually separated from the Yellow River. In the 17th century, the Middle River was excavated. In the 18th century, the Grand Canal, the Yellow River, and the Huaihe River

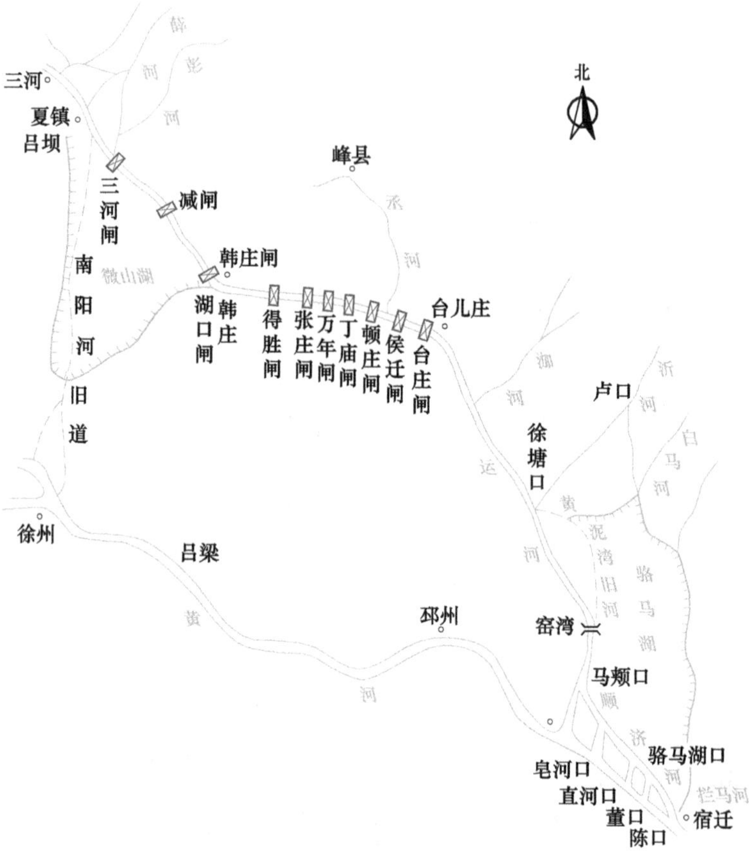

Figure 5.13: The Relationship between the Huitong Canal, the Middle Canal, and the Yellow River from the 17th Century to the 19th Century

con-verged at Qingkou in Huai'an, Jiangsu Province. Later, the Middle River and the Jia River were collectively known as the 'Middle Canal' (see Figure 5.13).

5.4 Sluices, Rivers, and Their Operation

The Huitong Canal extended south to Xuzhou, where it was connected with the Yellow River, and north to Linqing, where it was connected with the South Canal. The maximum elevation of the

Huitong Canal was at Nanwang Town in Jinan City, Shandong Province. Nanwang Town was also the divide between the Yellow River basin and the Huaihe River basin. At Nanwang Town, water in the Huitong Canal was diverted to the south and the north respectively. From the north of Nanwang Town to Linqing, water flowed from south to north, while from the south of Nanwang Town to Xuzhou, water flowed from north to south. The Huitong Canal was the reach with the largest elevation difference of the Grand Canal in the Yuan, Ming, and Qing dynasties. Before fossil-fuel-powered ships appeared, the only way for the ships to pass through the Huitong Canal was to take structural measures. The construction of continuous sluices in the Huitong Canal successfully solved the problem arising from the large elevation difference.

Because of the unique topography and water source conditions of the Huitong Canal, structural measures had to be taken to keep an appropriate water depth for navigation in the Huitong Canal. The operation of the continuous sluices in the Huitong Canal showed that the level of engineering planning, construction, and management of the Huitong Canal in the Yuan, Ming, and Qing dynasties had surpassed those of both the Huaiyang Canal and the Jiangnan Canal in the Song Dynasty, reaching the highest level in history thus far.

5.4.1 *Engineering System of Sluices and Rivers*

Continuous sluices are multiple sluices built in a canal. When the Huitong Canal was excavated in the 13th century, 20 sluices were constructed along it. Later, the number of the sluices were increased to 40. In the Qing Dynasty, there were 48 sluices on the Huitong Canal. Because the canal changed its course in different periods, the number of the sluices were increased or decreased to some extent. The sluices on the Huitong Canal were usually built of stone, although a small number were built of brick. Nowadays, the remains of the sluices are still usable. The construction ages and locations of the sluices on the Huitong Canal are shown in Table 5.1 and Figure 5.14.

Table 5.1: Control Sluices on the Huitong Canal in the Ming Dynasty (Late 16th Century)

No.	Location	Name	Distance from the upper sluice (km)	Year construction began: Yuan Dynasty	Year construction began: Ming Dynasty
1	Linqing, Shandong	Huitong		1293	
2		Linqing	0.5	1296	
3		Plate Sluice	1.5		1417
4		New Opened Sluice	0.75		1417
5	Qinping, Shandong	Daijiawan	15		1465
6	Tangyi, Shandong	Tuqiao	24		1471
7		Liangjiaxiang	25		1429
8	Liaocheng County, Shandong	Tongjiqiao Bridge	17		1418
9		Lihaiwu	10	1296	
10		Zhoujiadian	6	1300	
11	Yanggu County, Shandong	Qiji Lower	12	1297	
12		Qiji Upper	1	1295	
13		Acheng Lower	6	1299	
14		Acheng Upper	1.5	1298	
15		Jinmen Lower	5.	1299	
16		Jinmen Upper	1.5	1302	
17	Dongping County, Shandong	Daijiamiao	31.5		1537
18		Anshan	15	1289	
19		Jinjiakou	15		1525
20	Wenshang County, Shandong	Yuanjiakou	9		1506–1521
21		Kaihe	70	1335–1340	
22		Nanwang North (Shili)	7.5 metres to the north and south of the diversion of the water flow respectively		1465–1487
23		Nanwangnan (Liulin)			1465–1487

(*Continued*)

Table 5.1: (*Continued*)

No.	Location	Name	Distance from the upper sluice (km)	Year construction began: Yuan Dynasty	Year construction began: Ming Dynasty
24		Siqian	7.5		1506
25	Jining, Shandong	Division Gate (Upper)	55.5	1301	
26		Tianjing (Mid)	1.5	1284	
27		Zaicheng (Lower)	1	1284	
28		Zhaochun	3	1327	
29		Shifo	3.5	1319	
30		Xindian	9	1297	
31		Xin	4	1341	
32		Zhongjiaqian	2.5		1430
33		Shijiahzuang	3	1248	
34		Luqiao	2		1415
35		Zaolin	3	1318	
36	Yutai County, Shandong	Nanyang	6	1331	
37		Baliwan	4		1433
38		Mengyangbo	9	1303	
39	Peixian County, Shandong	Hulingcheng	4		1429
40		Gutou Upper	35	1315	
41		Gutou Mid	3		1484
42		Gutou Lower	4		1484
43		Xiegou	5		1433
44	Xuzhou, Shandong	Xinxing	9		1433
45		Huangjia	8		1458
46	Zaozhuang, Shandong	Xiazhen			1567

Sources: *Cao He Tu Zhi* by Wang Qiong, Ming Dynasty, *Ming Hui Dian* (Administrative Laws of the Ming Dynasty).

Figure 5.14: Key Projects on the Huitong Canal: Control Sluices and Water Tanks

The Huitong Canal ran through the uplands in Shandong Province. On the Huitong Canal there were 40 control sluices that were built during the period of the 13th century to the 14th century. The operational principle of the control sluices was similar to that of modern ship locks. Because of the sluices, a water surface with a gentle slope in the canal became a staircase-like water surface. In ancient times, when there was no fossil fuel as power, control sluices were important facilities for solving the problem of elevation differences.

Water tanks were artificial lakes on both banks of the canal. Water was stored in water tanks through diversion projects and then supplied to the canal through control sluices.

5.4.2 *Shape and Structure of Control Sluices*

The book *The History of the Yuan Dynasty: A Chronicle of Canals* describes the Huitong Sluice as follows:

> The sluice is 31.68 m long, 24.34 m wide, has two parallel walls on the main body of the sluice (each is 12.67 m long), two wing walls (each has a length of 10 m and a height of 6.34 m) and a gate with a length of 6.34 m. This is the dimensions of the then Huitong Sluice. Another type of sluice was called narrow sluice at the time, which was set up at a canal mouth to stop oversized ships from passing through it. A narrow sluice was the same as an ordinary control sluice except that it had a gate with a length of 3 m rather than 6.34 m.

The sluice consisted of piers, wing walls, a gate, and windlasses for the opening and closing of the gate. The water-facing sides of the piers and wing walls were made of boulder strips. Slots for the gate were set up on the main body of the sluice. The gate was made of stoplogs (see Figure 5.15). During the reign of the Qianlong Emperor in the Qing Dynasty, George Staunton, a Vice Ambassador of the British Diplomatic Corps to China, described in detail the structures of control sluices in the Huitong Canal:

> Like sluices in Europe, canal sluices in China have only one gate (not two gates: one is high and the other is low). The gate has a very simple structure and is easy to control. It does not take so much money to repair the gate. The gate was made of a few big planks that are set up in the slots of the levees on both banks of a river. The upper planks and lower planks are connected. When ships arrive at the sluice, the upper planks are lifted to have enough space for the ship to pass through.

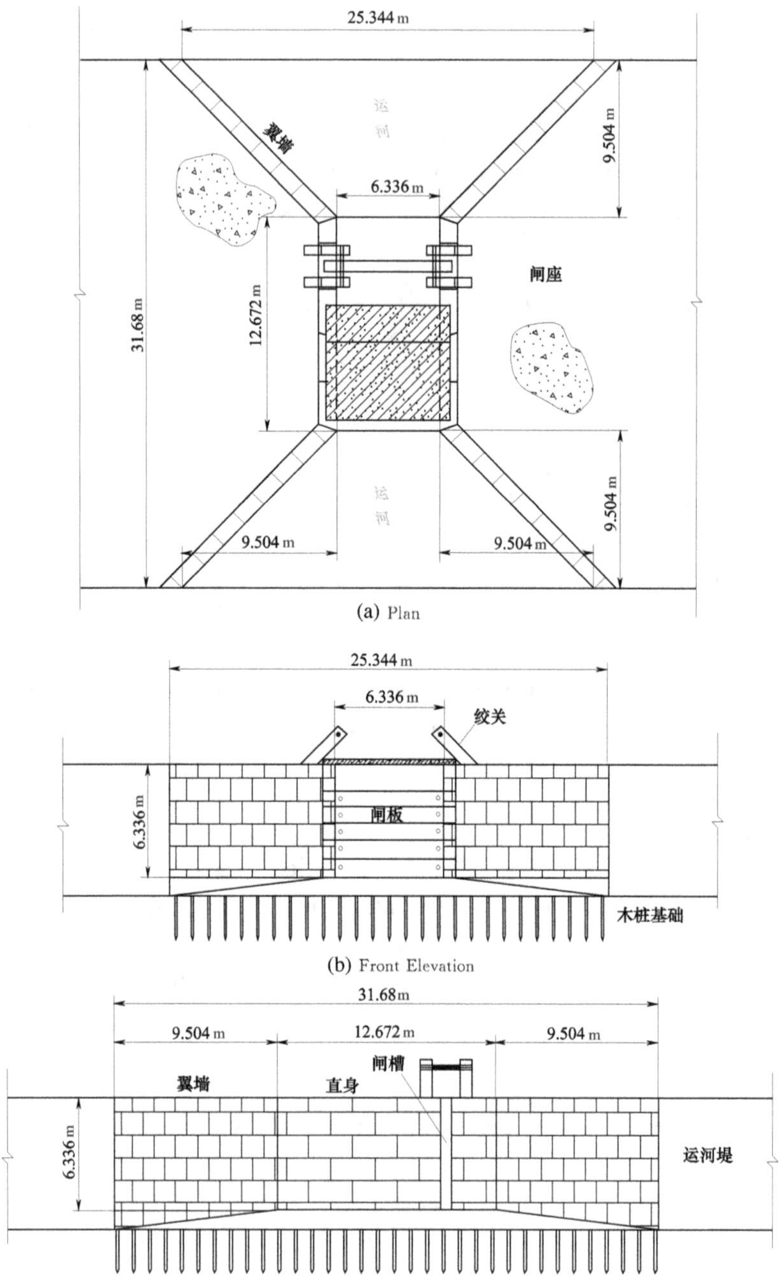

Figure 5.15: Shape and Structure of the Huitong Sluice

Figure 5.16: The Remains of the Liulin Sluice in the Huitong Canal

At the end of the Qing Dynasty, for the convenience of transportation, many control sluices were converted into two-gate sluices (see Figure 5.16).

5.4.3 *Sluices with Other Functions*

In literature, sluices on the Huitong Canal can be classified, according to their functions, as water storage sluices, water-intake sluices, sluices in barrages, overflow sluices, and so on. The sluices in barrages are regarded as control sluices of modern hydraulic structures. The functions and locations of other types of sluice can be identified through their names (see Figure 5.17).

5.4.4 *Use of Control Sluices on the Huitong Canal*

The control sluices on the Huitong Canal were aimed at controlling water volume and keeping an appropriate water depth for navigation. Because they were directly related to navigation, the sluices had a rigorous system for their opening and closing as well as maintenance.

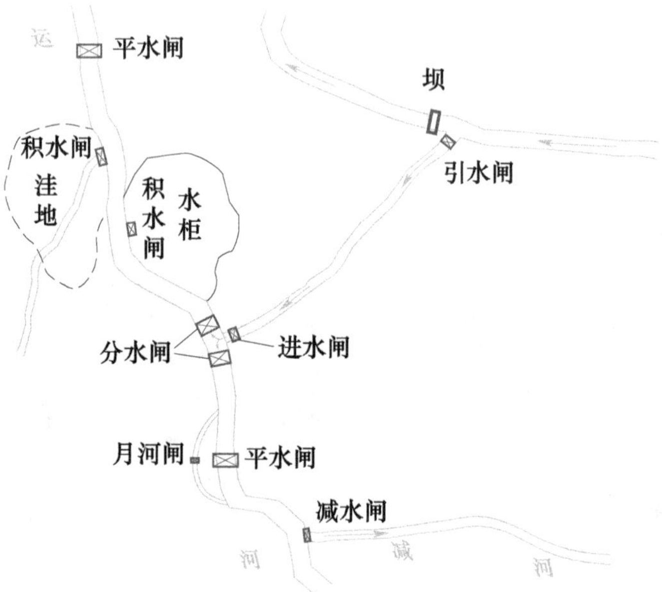

Figure 5.17: Sluices with Different Functions on the Huitong Canal

The sluices were managed with consideration of the following five aspects:

1. Conjunctive operation of adjacent upper and lower sluices. The principle was that when one gate was opened, the other was closed. According to historical records, when a ship reached a sluice, people on the ship had to receive signals from both of the upper and lower sluices and then one sluice was opened for the ship to pass through it. If there was adequate water in the canal, in order to improve the efficiency of navigation, the sluice manager could open or close the sluice according to specific situations. It was not permitted to open or close adjacent upper and lower sluices simultaneously, because doing so would lose too much water and result in a rapid fall in water levels.
2. Ships had to pass through a sluice in groups. When the number of ships in a group met a certain requirement, the gate could be opened. The purpose of doing this was to reduce the number of

times a sluice was opened and closed. Take Liulin Sluice as an example; it was specified that only when over 200 tribute rice ships arrived could the gate be opened. After the ships passed through the sluice, the gate was closed immediately. When enough water was stored to reach a certain water level, the sluice was opened in order for another group of boats to pass through it. Cracks between the gate stoplogs were stopped up by straw to reduce the leakage of water.

3. The water depth in the reach between two adjacent sluices was controlled strictly. The gate could not be opened if a specific water depth was not reached. In 1429, it was specified that except for ships paying tribute to the Emperor and emergency ships, for any other ships (such as those transporting tribute rice), official ships, civilian ships, and merchant ships, the gate could not be opened until the water level reached the sixth or seventh gate plank from the bottom.
4. During the low-flow period or at the end of each year, related sluices were closed and ships had to pass through a branch channel (or *Yue River* in Chinese) to make a detour. If necessary, a weir made of earth and straw was temporarily constructed to control the water level and ships were towed over the weir.
5. The cross-sections of the canal were designed specifically for navigation. In the Ming Dynasty, it was specified that the width of a ship for transporting tribute rice should be 3.5 m and that a fully loaded ship had a draught of 1.16 m. According to this specification, Wangong proposed that, as the dredging standard, the water depth in the canal should be no greater than 1.33 m and the width of the river channel should be no greater than 13.3 m (so two ships could sail in parallel in the canal) to reduce water use for navigation in the canal to the greatest extent possible.

In the 17th century, when he personally experienced the process of passing through the sluices in the Huitong Canal, George Staunton described it in detail:

> Substantial skills were required when ships passed through a sluice. A sailor stood on the bow of a ship with a big oar in his hand and

> commanded. The other passengers on the ship stood on both sides of the ship. The fenders on both sides of the ship were made of hide and stuffed with hair in order to avoid vibration caused by a collision between the ship and the stone of the sluice. The sluice was opened at fixed time every day. Every ship had to pay the shipping tax that was used to repair the sluice and levees.

Staunton was also impressed with change in water level of the canal and described it as follows:

> Every time the sluice was opened, the water discharged from the sluice to the downstream side was not great and, thus, the water level on the upstream side of the sluice fell only by a few inches. Water volume on the upstream side of the sluice reduced by the opening of the sluice could be compensated quickly by the water diverted from other rivers/lakes to the canal. Where there was a rapid water flow (namely the riverbed slope was steep) and the distance between two adjacent sluices was big, the opening of the sluice might result in water level falling by one or two feet. The canal was excavated along natural rivers [this is what the author imagined]. As a result, its depth was uneven and it was a meandering and wide channel. Generally, the number of sluices that a ship passed through within a day was not more than six.

It was the perfect engineering system of water sources, the systematic engineering system of control sluices, and the reasonable management system of water amount and level regulation that made the Huitong Canal maintain successful tribute rice transportation for 500 years. The adverse natural conditions of the Huitong Canal, its complicated engineering systems, and its sophisticated and rigorous operation and management have made it a canal unlike any other in the history of the world.

5.5 Key Projects on the Middle Canal

From 1522 to 1566, in the reach between Kaifeng and Xuzhou, the Yellow River burst its banks over and over again. When the Yellow

River burst its north bank, flood waters tended to flow northeast into the Huitong Canal. In 1565, the Nanyang New River was excavated to change the course of the reach of the Huitong Canal downstream of Nanyang (see Figure 5.12). In 1604, the Jia River was excavated. The course of the reach of the Huitong Canal downstream of Nayang was changed to the right bank of the Weishan Lake, so the Huitong Canal was connected with the Jia River, which flowed into the Yellow River at Suqian. After excavation of the Nanyang New River and the Jia River, the Grand Canal avoided Xuzhou-Hong and Luliang-Hong, but the reach of the Huitong Canal between Suqian and Huai'an was still the reach of the Yellow River. Because the canal mouth where the Jia River joined the Yellow River was silted up frequently, in 1686 the Middle Canal was excavated at Jinfu. The Middle Canal ran through the Luoma Lake, Suqian, and Taoyuan (modern-day Siyang County, Jiangsu Province), and ended at Zhongzhuang, Qinghe County. The Middle Canal joined the Yan River of Southern Jiangsu at Zhongzhuang. Ships in the Middle Canal could then reach the sea, as the Yan River flowed through Shanyang and into the sea at Andong. In 1703, a new canal mouth — Yangzhuang — was constructed. After that, heavily loaded ships going up for the transportation of tribute rice entered the Middle Canal through Yangzhuang Canal Mouth, while ships coming back from the north entered the canal through Zhongzhuang Sluice.

The Middle Canal joined the Huitong Canal at the outlet sluice of the Weishan Canal. It ran through Pei County in Jiangsu, Tongshan, Xinyi, Suqian, and Siyang and flowed into the Yellow River at Yang Zhuang Canal Mouth in Huai'an. It had a length of 190 km and an average slope of 13 per cent. The area that the Middle Canal ran through has an average annual precipitation of 700–800 mm. The precipitation in 3 months of the summer accounts for 40–50 per cent of that in a whole year. The precipitation has an even intra-annual distribution. Lakes such as the Weishan Lake, the Luoma Lake, and the Huangdun Lake were used as the water tanks of the Middle Canal. Through the regulation of the lakes, the water sources for the Middle Canal could be guaranteed.

5.5.1 *Embankment on the Middle Canal*

The reach of the Middle Canal downstream of Suqian was adjacent to the Yellow River. In some places, the distance between the Middle Canal and the Yellow River was even less than 50 m. The Huanglu Levee, downstream of the Shizi River at Suqian, was the south levee of the Middle Canal. In flood season, the canal was affected by floods from both the Yellow River and the Yi–Shu–Sishui rivers. Some reaches of the Middle River were either interrupted or blocked due to sedimentation.

After its excavation, the Middle Canal was the reach of the Grand Canal that faced the most serious flood-control problem. Downstream of Suqian, the purpose of the levee construction and flood control was to ensure the safety of the levee of the Middle Canal. On the bank of the Middle Canal that was closer to the Yellow River, there were many overflow weirs, water-release masonry sluices, and drainage culverts. Even nowadays, the water-release sluices are still magnificent. For example, Anjia Sluice (constructed in 1758) and Minbian Sluice (constructed in 1732) were both water-release sluices for the canal and sluices to divert water from the Huangdun Lake to the canal. Both of them were 30-m-wide masonry sluices with two gates. Each of the gates was 5 m wide. The pier in the middle was 5 m wide and 8 m high. Liulaojian Overflow Sluice had nine gates and a total length of 66 m. Each of its gates was 3 m wide and 6 m high.

5.5.2 *Hub Projects at the Mouth of the Shizihe River*

The reach of the Middle Canal near Pizhou was the lower Jia River and the lower Yi River, running through the Luoma Lake and the Huangdun Lake. At the outlet of the Huangdun Lake, the canal bifurcated and the branch channel was called Liutang River. The reach of the canal then ran parallel to the Yellow River (see Figure 5.18). The lake mouth of the Middle Canal was called 'Shizi River', and it was where the canal extended north to the Huangdun Lake and the Luoma Lake and south to the Yellow River. The area

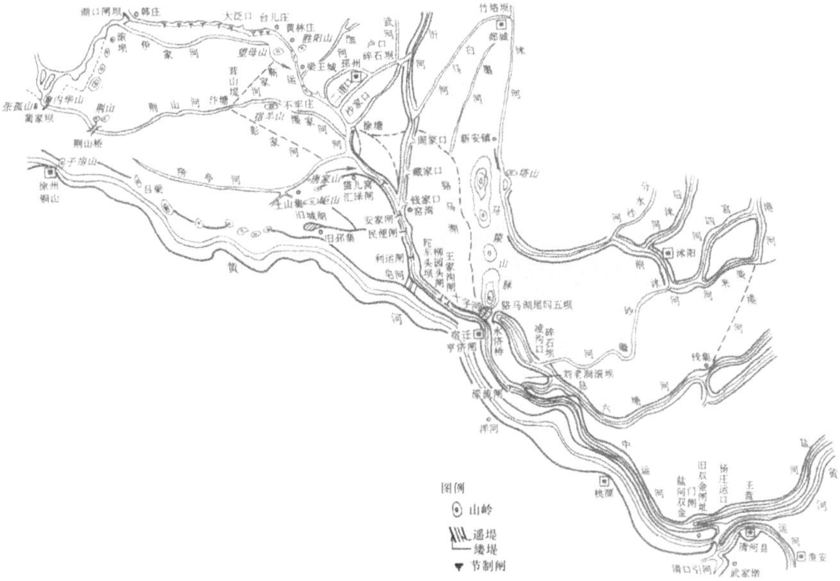

Figure 5.18: The Middle Canal and Its Key Projects in the 18th Century

around the Shizi River was the most important area. In 1689, the Middle Canal was excavated; the Zhuluo Dam was then constructed on the south bank of the Middle Canal near Huanglu Levee and a flood-release sluice was constructed to discharge floodwater from the Luoma Lake to the Liutang River. At the time, the hub project in the Shizi River was formed, which had the functions of diverting water from lakes to the canal, discharging floodwater from lakes to the Liutang River, and preventing the Yellow River from overflowing into the canal.

From the very beginning, the hub project in the Shizi River was a temporary project. When the water in the lakes could not meet the demand for water supply to the canal, Zhuluo Dam was abandoned to divert water from the Yellow River to the canal. When it became necessary to stop the floods of the Yellow River from flowing into the canal, Zhuluo Dam was rebuilt. The sluices around the lake mouth were either damaged by floods from the lakes or silted up by sediment from the Yellow River. The sluices and dams in the Shizi River were abandoned or rebuilt frequently. In short, during over 200 years of operation of the Middle Canal, the hub project in the Shizi River was one of the projects that had the highest maintenance cost.

The Liutang River and the Middle Canal came out of the Shizi River at the same place. The Liutang River bifurcated into two branches at the site of the Liulaojian Overflow Dam in Suqian; one flowed into the Middle Canal, while the other joined the Shu River in Andongna and then flowed into the sea. The Shajiang River, a tributary of the Shu River, was upstream of the Liulaojian Overflow Dam. Therefore, the Liulaojian Overflow Dam could regulate inflows from the Shajiang River and became a key project to supply water to the canal and discharge floodwater.

5.5.3 *The Zhongzhuang and Yangzhuang Canal Mouths*

In the Kangxi Era during the Qing Dynasty, the Middle Canal was excavated and the canal mouth was located in Zhongjiazhuang, Qinghe. As a result, it was named as the 'Zhongzhuang Canal Mouth'. At the canal mouth, the Yellow River, the Huaihe River, and the Middle Canal converged, and there were rapids there. It was difficult for heavily loaded ships to go up and pass through the canal mouth. Twelve years after the construction of the Zhongzhuang Canal Mouth, a new canal mouth — named 'Yangzhuang' — was constructed downstream of Yangzhuang, Shuangjin Sluice was built and a branch channel was excavated. The branch channel was on the floodplain, with length of 3.5 km. The channel flowed into the Yangzhuang Canal Mouth and then joined the Middle Canal, which avoided ships sailing through the rapids at the confluence of the Yellow River, the Huaihe River, and the Middle Canal. Since then, ships for the transportation of tribute rice had been able to sail downstream with the current in the Yellow River and then enter the Middle Canal at the Yangzhuang Canal mouth as soon as they left Qingkou. As a result, people could avoid pulling heavily loaded ships against the current in the Yellow River for hundreds of kilometres (see Figure 5.19).

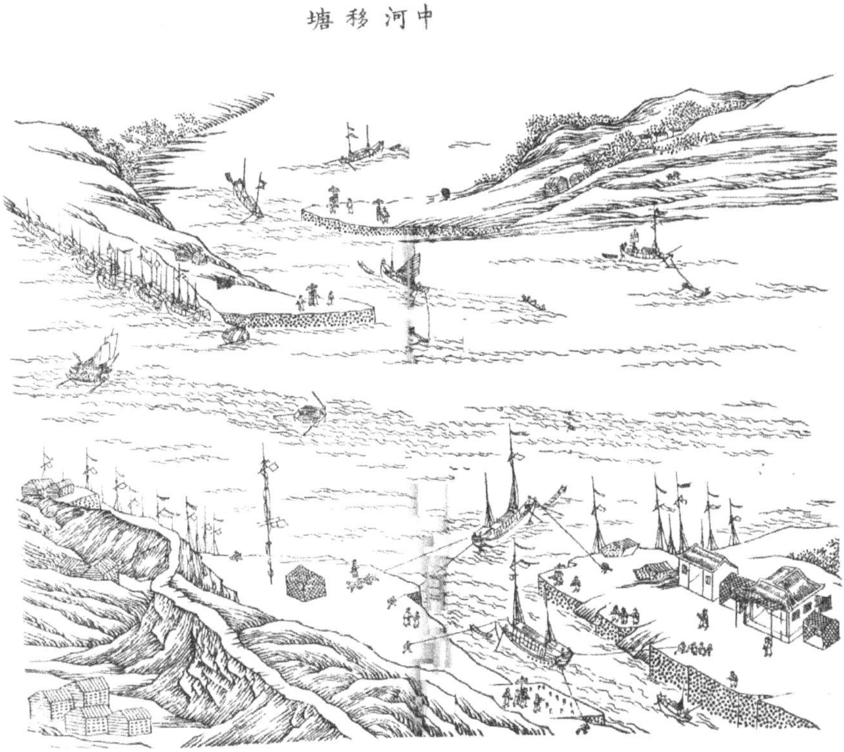

Figure 5.19: The Canal Mouth at the Confluence of the Yellow River, the Huaihe River, and the Middle Canal

Note: This picture is from Linqing's *Pictures in Heavy Snows* (Qing Dynasty). The scene shows that a fleet of ships went up from the Huaiyang Canal to the Middle Canal to sail across the Yellow River between the 17th and 19th centuries.

Chapter 6

The Huaiyang Canal

6.1 Introduction

Originating from the Han Ditch excavated by King Fu Chai of the Wu Kingdom in 486 BC, the Huaiyang Canal connects the Yangtze River and the Huaihe River and is the oldest canal with a definite record of its excavation period. The canal goes from Huai'an to Yangzhou across numerous lakes; as a result, it was historically known as the 'Lake Channel'. After the Yellow River changed its course so that the lower reach of the Huaihe River became part of the Yellow River in the Southern Song Dynasty, Qingkou, in Huai'an (or the north mouth of the Huaiyang Canal), was the confluence of the Yellow River, the Huaihe River, and the Huaiyang Canal. Because of the sedimentation of the Yellow River, the riverbed elevation of the lower reach of the Huaihe River was gradually raised, which had a large impact on the water system of the lower Huaihe River and resulted in the formation of the Hongze Lake, the fifth largest freshwater lake in China. It also meant that the Huaihe River changed its course and flowed into the Yangtze River. In the Ming and Qing dynasties, in order to ensure that the canal was suitable for the transportation of tribute rice, almost all types of water projects were constructed around Qingkou, among which the levee of the Hongze Lake was made of masonry. Since then, the Hongze Lake has become an artificial reservoir with an important status in the world history of civil engineering.

6.2 Basic Conditions

The Huaiyang Canal runs through Huai'an and Yangzhou, two prefecture-level cities in Jiangsu Province, with a total length of 160 km. It is located in a low-lying area with a dense network of rivers and lakes and relatively rich water resources.

Originating from the Han Ditch excavated in the Spring and Autumn Period, the Huaiyang Canal has experienced 2,500 years of development and is now still playing an important role in navigation.

6.2.1 *Overview of the Natural Conditions of the Huaiyang Canal*

The Huaiyang Canal is located in a region where the elevation is higher in the west than that in the east and slightly higher in the north than that in the south, with riverbed gradients from 0.033 to 0.1 per cent. The region to the west of the Huaiyang Canal is called the Yunxihuqu Plain and the region to the east is the Lixiahe Plain. The elevations of the bottoms of the lakes in the west of Huaiyang Canal are between 3 and 5 m and those of the shores and terraces of the lakes are between 6 and 10 m. A few monadnocks, such as the Shu Hillock in Yangzhou and the Tian Hill in Gaoyou, have an elevation of approximately 50 m. The Lixiahe Plain was originally the remnant of a bay and was later formed by the sediments from the Yangtze River, the Huaihe River, and the Yellow River.

The Huaiyang region has a subtropical monsoon climate with an average temperature between 14°C and 15°C . Annual average precipitation ranges from 800 to 1,100 mm in a north-to-south direction. The maximum monthly rainfall generally occurs during the plum rain season in June or during the typhoon season from August to September.

6.2.2 *The Historical Evolution of the Huaiyang Canal*

Originating from the Han Ditch excavated by the Wu Kingdom of the Spring and Autumn Period in 486 BC, the Huaiyang Canal has a history that stretches for 2,500 years. At the very beginning, it was an artificial channel partially connected to natural rivers and lakes. With the improvement in the past dynasties, in the Song Dynasty it became a relatively independent canal system controlled by water projects. In the Yuan, Ming, and Qing dynasties, because the Yellow River claimed the Huaihe River, the area around the north mouth of the Huaiyang Canal — namely Qingkou, as the conflu'ence of the Yellow River, the Huaihe River, and the Huaiyang Canal — became an important region in terms of the canal's improvement. The canal was kept navigable through a large number of structural measures, as well as sophisticated operation and management. Due to the sediment from the Yellow River, the Huaihe River eventually changed its course and flowed into the Yangtze River at the end of the Qing Dynasty. Since the turn of the 20th century, through some improvement measures such as diverting water from the Huaihe River to the Yangtze River, the Huaiyang Canal has been separated from the channels by which the Huaihe River is connected to the Yangtze River, and it has played an important role in navigation and the South–North Water Diversion Project.

6.2.2.1 *Excavation of the Han Ditch in the Spring and Autumn Period*

In the Spring and Autumn Period, in order to dominate the Central Plain (the region around the middle and lower reaches of the Yellow River), the Wu Kingdom excavated the Han Ditch to connect the Yangtze River and the Huaihe River, which was used to transport grain and its troops to the north. The Han Ditch (see Figure 6.1) is the origin of the Huaiyang Canal. The excavation of the canal in the ancient times made full use of the natural rivers and lakes. Some artificial channels were excavated to connect the rivers and lakes. The then-Han Ditch diverted water from the Yangtze River near Yangzhou, went up, passed through Gaoyou, turned northeast into the Sheyang Lake, turned northwest after running out of the Sheyang Lake, and finally flowed into the Huaihe River at Mokou, which is north of Huai'an (see Figure 6.2). It took 2 years to excavate the Han Ditch successfully. The Navy of the Wu Kingdom went up from the Yangtze River, passed through the Han Ditch, entered the Si River and the Yi River via the Huaihe River, and defeated the Qi Kingdom and the Jin Kingdom. As a result, the Wu Kingdom became the overlord. Since then, the Han Ditch has been an important channel connecting the Yangtze River and the Huaihe River.

Figure 6.1: The Remains of the Han Ditch (2013)

Note: With the improvement of the past dynasties, the Han Ditch became an important canal connecting the Yellow River, the Huaihe River, and the Yangtze River, while at the same time having a large impact on the environmental evolution of the Huaiyang region. Nowadays, the relics of the Han Ditch in Yangzhou City are an integral part of the city's river system and also an urban landscape river with historical and cultural value.

6.2.2.2 *The Overall Improvement of the Huaiyang Canal in the Sui and Tang Dynasties*

During the Sui Dynasty, the Huaiyang Canal was called 'Shanyangdu'. In AD 587, in order to unify the Jiangnan region, the Han Ditch was improved along Chen Deng's route. After unifying the whole country, Emperor Yang of the Sui Dynasty excavated the Ji Channel, while at the same time mobilising over 100,000 people in Huainan to renovate and widen the Huaiyang Canal. Between the south of Shanyang County and the Yangtze River, the Huaiyang Canal had a length of over 150 km and a width of approximately 60 m, which was wide enough for warships to pass through. Along the two banks of the Huaiyang Canal, royal roads were constructed and willows were planted. The southern end of the Huaiyang Canal was no longer

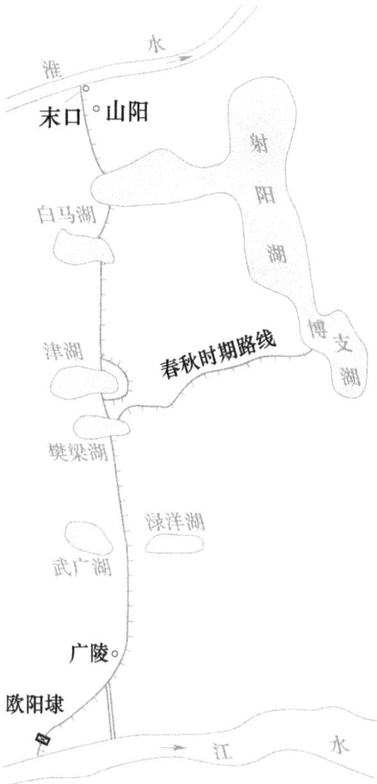

Figure 6.2: Schematic Diagram of Improvement of the Han Ditch

Note: In the Han Dynasty, the Han Ditch was also called Qushui or Zhongdushui. The main lakes along the Han Ditch included the Sheyang Lake, the Baima Lake, the Jin Lake, the Fanliang Lake, the Wuguang Lake and the Luyang Lake. From AD 196 to 220, Chen Deng, Mayor of Guangling (today's Yangzhou) straightened the Han Ditch: from the south of the Fanliang Lake to the Yangtze River, the Han Ditch was kept unchanged; a channel was excavated between the north of the Fanliang Lake (25 km northwest of today's Gaoyou County) and the Jin Lake, and another channel was excavated between the north of the Jin Lake and the Baima Lake (7.5 km northwest of today's Baoying County); then the canal passed through the west of Shanyang City (today's Huai'an) and entered the Huaihe River at Shanyangkou. From AD 363 to 365, the Jin Lake was more than 5.5 km wide, so the wind howled and the waves rose. As a result, between the south and the north of the Jin Lake, a channel with a length of over 10 km was excavated along the eastern shore of the Jin Lake to make a detour around the Jin Lake and avoid the wind and the waves.

connected to the Yangtze River via Ouyangdi but was instead directly connected to the Yangtze River at Yangtze, east of Ouyangdi (near the Yangtze Bridge, 7.5 km south of modern-day Yangzhou). At that time, the route of the Huaiyang Canal was basically determined.

Figure 6.3: Schematic Diagram of the Projects in the Yilou Canal

Note: To the south of Yangtze, there was a sandbank in the Yangtze River named 'Guazhou'. Due to the sedimentation in the Yangtze River, Guazhou was gradually connected to the northern bank of the Yangtze River. The mouth of the Jiangnan Canal at Jingkou was originally opposite Yangtze, but nowadays ships from the Jiangnan Canal have to go west and make a detour, traveling more than 30 km to reach Yangtze and then entering the Huaiyang Canal. In 738, Qi Huan, the prefectural governor of Runzhou, excavated a canal on Guazhou that was named the 'Yilou Canal'. It had a length of 12.5 km. Along the Yilou Canal, some control projects were constructed such as the Yilou Weir and the Erdoumen Gate. The Erdoumen Gate was the first recorded ship lock in the world. Ships travelled 10 km along the Yangtze River from Jingkou to Guazhou and then entered the Yilou Canal.

In the Tang Dynasty, the Huaiyang Canal was called the 'Yangchu Canal'. The multi-purpose water projects at the canal mouth connected to the Yangtze River were initially formed in the Tang Dynasty. The canal mouth project at Guazhou and the canal mouth project at Yizheng were constructed during that period (see Figure 6.3). Ships from the upper Yangtze River could enter the canal through the canal mouth at Yizheng, while ships from the Jiangnan Canal could enter through the canal mouth at Guazhou. Due to a relatively high elevation of the area around Yangzhou, water-regulating ponds such as the Chengnong Pond, Shanglei Pond, Xialei Pond, Goucheng Pond, and Xiaoxin Pond were required. They were commonly known as the 'Five Ponds in Yangzhou'.

6.2.2.3 *The Perfection of the Project System of the Huaiyang Canal in the Song Dynasty*

During the Song Dynasty, economic emphasis was shifted to the south and the region around Jiangsu and Zhejiang became the main area for the production of grain. In the Northern Song Dynasty, the annual amount of tribute rice transported from Jiangsu and Zhejiang to Bianjing (modern-day Kaifeng) was 6 million dan, reaching 8.5 million dan at most. An increase in the transportation of tribute rice resulted in a higher demand for the project system of the canal and also greatly promoted the development of the canal in the Song Dynasty. During that period, the Guishan Canal was excavated to bypass the Huaihe River, the levee between the canal and the lakes (such as the Shaobo Lake in Gaoyou) was constructed, and the flood-control projects on the canal were built. Such projects made the Huaiyang Canal a water project system with more independent channels and better project facilities.

The Tongji Canal, which was excavated in the Sui Dynasty and called 'Bian Canal' in the Song Dynasty, entered the Huaihe River at Xuyi, while the Huaiyang Canal entered the Huaihe River at Huaiyin. Between Xuyi and Huaiyin, the main stream of the Huaihe River was used for navigation, where ships often capsized. In order to avoid the danger of sailing in the Huaihe River, the Sha Canal, the Hongze Canal, and the Guishan Canal were excavated successively near the southern bank of the Huaihe River in the Song Dynasty. Later, all of the three canals were generally known as the 'Guishan Canal' (see Figure 6.4). It took 100 years in total to complete the construction of the artificial canal between Huai'an and Xuyi. Since then, ships have been able to directly reach Xuyi via the Huaiyang Canal, then cross the Yangtze River and enter the Bian Canal.

There were many lakes along the Huaiyang Canal, and ships sailed in the lakes. Most of the reaches of the canal did not have the shape

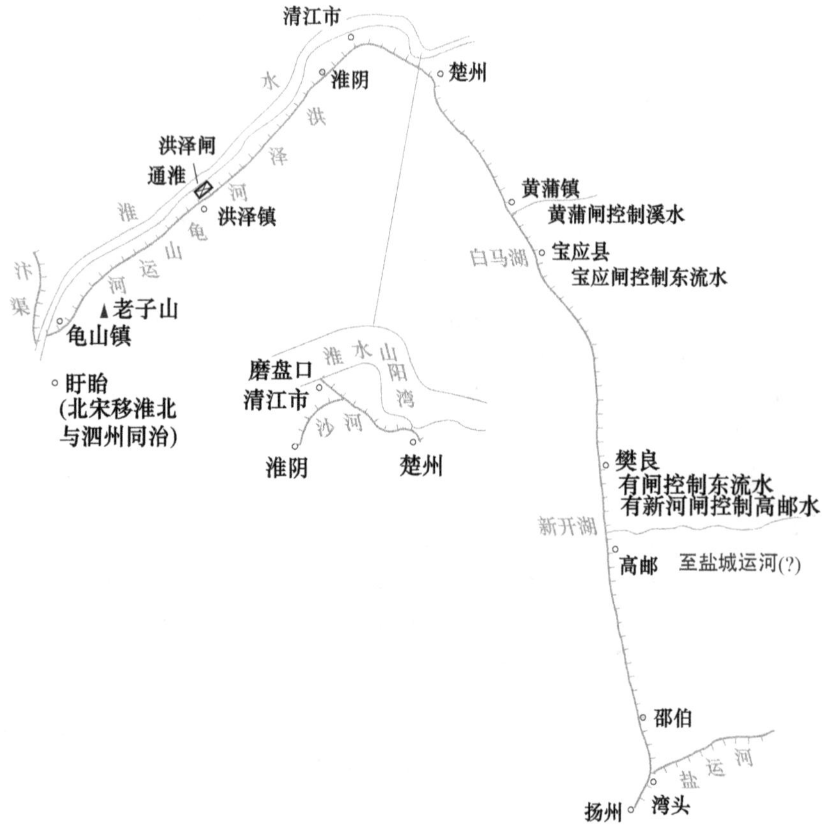

Figure 6.4: Schematic Diagram of the Use of the Canal in the Song Dynasty to Avoid Sailing in the Huaihe River

Note: 1. In 984, Qiao Weiyue, a transportation official in Huainan, excavated a canal that started at Mokou in the north of Huaiyin and entered the Huaihe River at Mopankou (near modern-day Matou Town) in Huaiyin. The canal was named as the 'Sha Canal' and had a length of over 20 km. The completion of the Sha Canal meant the danger of sailing in the Huaihe River could be avoided.

2. From 1049 to 1054, Xu Yuan, an official in charge of the tribute rice transportation in Jianghuai, extended the Sha Canal by excavating a canal between Mopankou and Hongze. It was called the 'Hongze Canal' and had a length of 24.5 km.

3. In 1083, after a suggestion from Luo Zheng and Jiang Zhiqi, the Hongze Canal was extended west to Guishan. Chen Youfu, an official in charge of canals, was responsible for the planning and implementation of the canal. One hundred thousand people were mobilised to excavate the canal with a length of approximately 28.5 km. It took over one month to complete the canal.

of a river channel. Some lakes had an extensive water surface in which strong winds and huge waves often occurred, making it dangerous for ships to sail there. The construction of the levees between the canal and the lakes started in the Song Dynasty, which gradually formed a levee system and made the canal independent of natural rivers and lakes, thus greatly enhancing the human capacity to control the canal. According to records, from 1004 to1007, in 1025, and in 1194, the canal levee between Huaiyin and Jiangdu was constructed, which not only made the navigation safer, but also created conditions for the reclamation of the area around the canal. In order to ensure the safety of the canal during the flood season, a large number of flood-discharging projects were constructed on the eastern levee of the canal, such as overflow weirs and sluices for drainage. There were nearly 100 flood control projects at their peak. The projects discharged water to the Lixiahe Plain when large floods occurred in the canal, which ensured the safety of the canal while at the same time exacerbating the flood and water-logging disasters in the Lixiahe Plain.

6.2.2.4 *The Huaiyang Canal in the Yuan and Ming Dynasties, against a Background of the Yellow River and the Huaihe River Changing Their Courses*

In 1128, the Yellow River changed its course and ran south, finally claiming the lower reach of the Huaihe River. The reaches shared by the Yellow River and the Huaihe River were gradually silted up. Qingkou in Huai'an became a confluence of the Yellow River, the Huaihe River, and the Huaiyang Canal. The area around the confluence was the emphasis of the canal improvement. In order to ensure a smooth transportation of the tribute rice, a general plan was enforced to store clear water and flush out sediments, namely to heighten the Gaojia Weir so as to store clear water in the Hongze Lake and then use the water stored in the Hongze Lake to flush out riverbed sediments from the reaches shared by the Yellow River and the Huaihe River. Meanwhile, through construction of a large number of water projects and sophisticated transportation management, it was ensured that boats transporting tribute rice could pass through Qingkou smoothly.

The Huaiyang Canal was commonly known as the 'Lake Channel' as it had a close relationship with lakes west of it. Each lake provided a water source for the canal. Meanwhile, the construction of the canal impeded the discharge of floods to the east. In the Ming Dynasty, the levee system was further perfected on the basis of that of the Song Dynasty. Levees in some important reaches were reconstructed as brick levees or stone levees. Meanwhile, flood-discharging sluices and weirs were constructed on the levees to ensure the safety of flood control. The independence of the canal project system and the human capacity for controlling the canal were enhanced. Among all the lakes, the Baoying Lake and the Siguang Lake had the most significant flood hazard, where the canal levees often broke. The main measures to deal with the hazard included strengthening and heightening the levee, excavating the Yue Canal, constructing sluices between the canal and the lakes, and building flood-discharging sluices on the eastern levee of the canal. A complete levee and flood-control project system was basically established and lasted until the Qing Dynasty (see Figure 6.5).

Figure 6.5: Schematic Diagram of the Huaiyang Canal System and its Related Projects

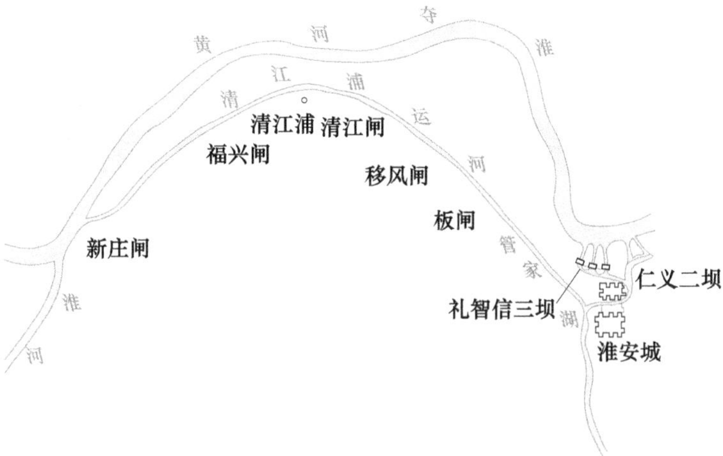

Figure 6.6: Schematic Diagram of Five Dams in Huai'an and Five Sluices Along the Qingjiangpu Canal

Note: At the beginning of the Ming Dynasty, five dams — Ren, Yi, Li, Zhi, and Xin — were used to make boats able to reach the Huaihe River. Boats transporting tribute rice entered the Huaihe River through the Ren and Yi Dams, while civilian and merchant boats entered the Huaihe River through the Li, Zhi and Xin Dams. The boats were towed over the dams. The dams were constructed by use of such materials as earth and tree branches so as to reduce abrasion to the boats; they were therefore called 'soft dams'. In AD 1415, Chen Xuan dredged the Sha Canal excavated by Qiao Weiyue of the Song Dynasty and renamed it 'Qingjiangpu'. The Qingjiangpu Canal started at the Guanjia Lake west of Huai'an and ended at Yachenkou, through which boats went into the Huaihe River. Along the Qingjiangpu Canal, five sluices — Xinzhuang, Fuxing, Qingjiang, Yifeng, and Ban — were constructed to regulate water flow. During that period, the Qingjiangpu Canal became the main channel connecting the Huaiyang Canal and the reach shared by the Huaihe River and the Yellow River.

In the early Ming Dynasty, the Qingjiangpu Canal at the northern end of the Huaiyang Canal was used to allow boats to cross the Huaihe River (see Figure 6.6). In the late Ming Dynasty, Pan Jixun put into practice the river-training strategy of 'storing clear water and flushing out sediments' so as to improve the rivers and protect the canal. The levees in the lower reach of the Yellow River and the Gaojia Weir were constructed. Qingkou was the confluence of the Yellow River, the Huaihe River, and the Huaiyang Canal, so all the projects were located around Qingkou. As the riverbed of the reach shared by the Yellow River and the Huaihe River was raised due to sedimentation, the riverbed elevation

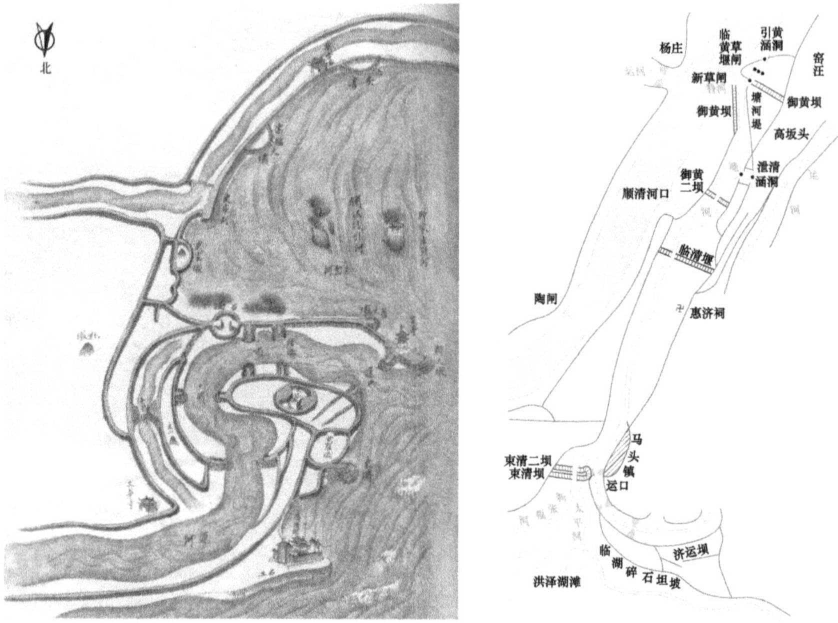

(a) During the reign of Emperor Kangxi (b) During the reign of Emperor Daoguang

Figure 6.7: Schematic Diagram of Project System at Qingkou and Artificial Ponds Used to Supply Water for the Canal

difference between the Huaiyang Canal and the reach shared by the Yellow River and the Huaihe River became increasingly large. With a limited space, the canal had to be extended and many sluices had to be constructed to regulate water flow in order to ensure that boats transporting tribute rice could pass through the canal smoothly. The effects of the projects, however, were not as good as expected (see Figure 6.7).

In the Qing Dynasty, a lot of manpower, financial resources, and material resources were put into the construction of projects at Qingkou, but these could not change the facts that the Yellow River continued to be silted up, sediment-laden water flowed backward into the Hongze Lake, and the transportation of tribute rice became increasingly difficult. In AD 1851, the Huaihe River changed its course at the Gaojia Weir and entered the Yangtze River via the Huaiyang Canal and the lakes along it. In 1855, the bank of the Yellow River broke at Tongwaxiang in Lanyang, Henan Province, and changed its course, claiming the Daqing River and then flowing into the sea. Before long, the government of the Qing Dynasty announced that transportation of tribute rice would be stopped; the Beijing–Hangzhou Canal, as a traffic artery in Eastern China, stepped down from the stage of history. The projects in the canal were formed out of unified management, and the canal gradually evolved into separated local channels.

6.2.2.5 *Improvement of the Huaiyang Canal in Modern Times*

After the transportation of tribute rice was stopped, the navigational function of the Huaiyang Canal was not abandoned. In 1936, the government of the Republic of China enforced the planning of the 'Huaihe River training'; the channels connecting the Huaihe River and the Yangtze River were improved, levees were reinforced, and river channels were dredged. While enforcing the planning of the Huaihe River training, the navigation in the canal was taken into consideration, with three ship locks named 'Huaiyin', 'Shaobo', and 'Gaoyou' constructed along it (see Figure 6.8). After 1949, due to systematic improvement, channels connecting the Huaihe River and the Yangtze River were separated from the canal. Nowadays, the Huaiyang Canal is a second-grade waterway for navigation, while at the same time serving as the water-diversion channel of the eastern route of the South-to-North Water Diversion Project.

Figure 6.8: The Shaobo Ship Lock with Two Lanes (2009)

Note: The Shaobo Ship Lock has a width of 10 m and a net length of 100 m with a maximum waterhead of 7.7 m. Its construction began in March 1934 and was completed in June 1936. The construction of the Huaiyin Ship Lock coincided with that of the Shaobo Ship Lock, and the design of its structure was the same as that of the Shaobo Ship Lock — with a maximum waterhead of 9.2 m between the upstream and downstream sides. The chamber of the Gaoyou Ship Lock is 10 m wide and 30 m long with a single steel gate. The width of its entrance and exit is 5.8 m. Its construction started in June 1935 and was completed in April 1936.

6.3 Key Projects

The Huaiyang region is characterised by a dense river network and is abundant in water resources. Its terrain is high in the west and low in the east. The Huaiyang Canal (see Figure 6.9) traverses the Huaiyang region in a north-to-south direction. It connects lakes along it, while at the same time impeding the discharge of local floods to the east, which represents the core natural background and the features of the water system. The control projects of the Huaiyang Canal are mostly concentrated in places where the canal intersects natural rivers and lakes horizontally. The areas around the mouths of the Huaiyang Canal where the canal intersects the Huaihe River and the Yangtze River horizontally were the ones with the densest concentration of projects. Among those areas, the area around Qingkou was the one with the densest concentration of water projects along China's Grand Canal during the Ming and Qing dynasties. The Gaojia Weir, which was constructed to systematically improve the Yellow River, the Huaihe River, and the Huaiyang Canal, was a masonry dam that formed an artificial reservoir — the Hongze Lake, which had a storage capacity of 13.5 billion m^3. The ship lock constructed in the

Figure 6.9: The Reach of the Huaiyang Canal in Yangzhou (2009)

Huaiyang Canal in the Tang Dynasty was the world's first in the contemporary sense, representing the highest scientific achievement in navigational engineering at the time.

6.3.1 *Channels and Water System Features on the Huaiyang Canal*

The Huaiyang Canal was historically known as the 'Lake Channel', which fully illustrates the water system features of the canal. The Huaiyang region is characterised by a low-lying terrain with many natural lakes. The Huaiyang Canal made full use of natural waters and water resources. Initially, in order to connect the Yangtze River and the Huaihe River, a continuous waterway was formed through some channels that connected adjacent lakes. After the Song Dynasty, the levees between the Huaiyang Canal and lakes were constructed one after another so as to reduce the risks associated with strong winds and huge waves in the lakes. The lakes became the regulating pools for the canal.

After 1128, the Yellow River claimed the Huaihe River, which had a far-reaching impact on all water systems in the Huaiyang region, including the canal water system. The Yellow River carried floods in the Huaihe River to the Huaiyang region and seriously affected the region, which became the main reason for disasters there after the Song Dynasty, especially during the Ming and Qing dynasties. In the Ming Dynasty, the Gaojia Weir was constructed, which resulted in the formation of the Hongze Lake. The overflow dam on the Gaojia Weir was the main reason for floods in the Lixiahe region. The terrain of the Huaiyang region is high in the west and low in the east, and the Huaiyang Canal traverses the region in a north-to-south direction. In addition, after the Song Dynasty, a levee system was established, impeding the discharge of floods and thus increasing flood hazards. This is an irreconcilable contradiction caused by regional disasters in the history. It was not until the Huaiyang Canal and the channels connecting the Huaihe River and the Yangtze River were separated after 1949 that the situation of serious flooding was eased.

6.3.2 *Control Projects*

Generally speaking, the Huaiyang Canal has a relatively gentle riverbed gradient. The reach between Gaoyou and the Yangtze River has a slightly steep riverbed gradient. After the Yellow River claimed the Huaihe River, the elevation difference of the land surface around Huai'an gradually increased. As a result, control projects were concentrated on the reaches near the northern and southern mouths of the canal as well as the reaches between Gaoyou and Shaobo. This section will discuss the historical evolution of the control projects and their features in terms of project types.

The earliest control projects on the canal are weirs and dams. Along the Huaiyang Canal, weirs were constructed near the canal mouths connected to the Yangtze River and around Shaobo. The recorded weirs in the Huaiyang Canal include the Ouyang Weir, the Shaobo Weir, the Qinliang Weir, the Sanmei Weir, and the Jingliang Weir. Their function was mainly to reduce the impact of frequent changes in water levels on the canal, balancing the difference between adjacent reaches. Boats were towed over the weirs. 'The Directory of International Canal Relics' says:

> A long time ago, Chinese people realised that if a slope has a moderate gradient, flatboats can be towed up the slope. Based on this principle, they invented weirs with slopes on their upstream side and downstream side. Such weirs are the ones in the canal. From the Eastern Jin Dynasty to the Sui and Tang dynasties, weirs and dams were always the main type of control projects in the Huaiyang Canal. After the Tang Dynasty, there were already developments in sluice technology, but weirs and dams were still widely used until the Song Dynasty. At the beginning of the Song Dynasty, there were five weirs in the Huaiyang Canal, which were named Longzhou, Xinxing, Zhuyu, Shaobo and Beishen respectively and located in the southern reach of the Huaiyang Canal.

In the 10th century, the world's first ship lock in the contemporary sense was constructed on China's Grand Canal, and it was located at the northern end of the Huaiyang Canal. In AD 984, Qiao

Weiyue, a transportation official in Huainan, rebuilt the Xihe Weir in the Sha Canal into two sluices with a distance of 73.5 m between them. The gates of the sluices were wood stoplogs, which were opened and closed by windlasses. Their method of operation is the same as that of modern ship locks. These kind of ship locks are considered the oldest ones in literature, and at the time they were referred to as multi-sluice projects. After reconstruction, there was no need to tow boats over weirs, which greatly enhanced the efficiency of navigation. In AD 1022, the Zhenzhou Sluice at the canal mouth connected to the Yangtze River was also rebuilt into a multi-sluice project. In the Ming and Qing dynasties, multiple ship locks at the Qingkou canal mouth were also set up to avoid a huge waterhead between the Huaiyang Canal and the Yellow River.

6.3.3 *Multi-purpose Projects at the Mouths of the Canal*

The projects at the mouths of the Huaiyang Canal refer mainly to the multi-purpose projects at its northern and southern ends, which were used to connect the Huaiyang Canal to the Huaihe River and the Yangtze River. The densest concentration of projects was around the mouths of the Huaiyang Canal. After the Yellow River claimed the Huaihe River, the canal mouth at the northern end of the Huaiyang Canal, namely Qingkou, not only became the confluence of the Yellow River, the Huaihe River, and the Huaiyang Canal, but was also the core of the improvement of the Beijing–Hangzhou Canal and even the water projects across the nation. In the Ming and Qing dynasties, a huge investment was made in the projects at Qingkou so that it became the place with the largest number of multi-purpose projects along the Grand Canal. Among these multi-purpose projects were the then-largest masonry weir, the Gaojia Weir, and the largest artificial reservoir, the Hongze Lake.

6.3.3.1 *The Northern Canal Mouth — Qingkou*

Originally, 'Qingkou' referred to the mouth of the Sishui River (also called the Southern Qing River) that was connected to the Huaihe River. After the Yellow River flowed into the Huaihe River by claiming the Sishui River, 'Qingkou' referred to the site at which the (clear) water of the Huaihe River flowed into the Yellow River (with turbid water). Qingkou was the confluence of the Yellow River, the Huaihe River, and the Huaiyang Canal. The key issue of the projects at Qingkou was that the sediment from the Yellow River were deposited in the riverbed, so it became difficult for the Huaihe River to discharge water and the waterhead between the Huaihe River and the Huaiyang Canal gradually increased. In addition, the levees of the Yellow River and the Huaihe River often broke and the Yellow River caused water to flow backward, which resulted in sedimentation in Qingkou and the Huaiyang Canal. As a result, the water flows around Qingkou were extremely unstable, which had a large impact on the smooth navigation of the Huaiyang Canal. In the Ming and Qing dynasties, in order to improve rivers and protect the canal, as well as to make overall arrangements for the Yellow River, the Huaihe River, and the Huaiyang Canal, a large number of projects were constructed at Qingkou. Generally speaking, the projects fell into three categories — storing clear water to flush out sediments (avoiding using water from the Yellow River and diverting water from the Huaihe River), impounding water in water-regulating ponds, and supplying water to ensure there was enough water in the canal (see Figure 6.10).

The Gaojia Weir was an integral part of the multi-purpose projects at Qingkou (see Figures 6.11 and 6.15) and also the key project of

Figure 6.10: Qingjiang Great Sluice

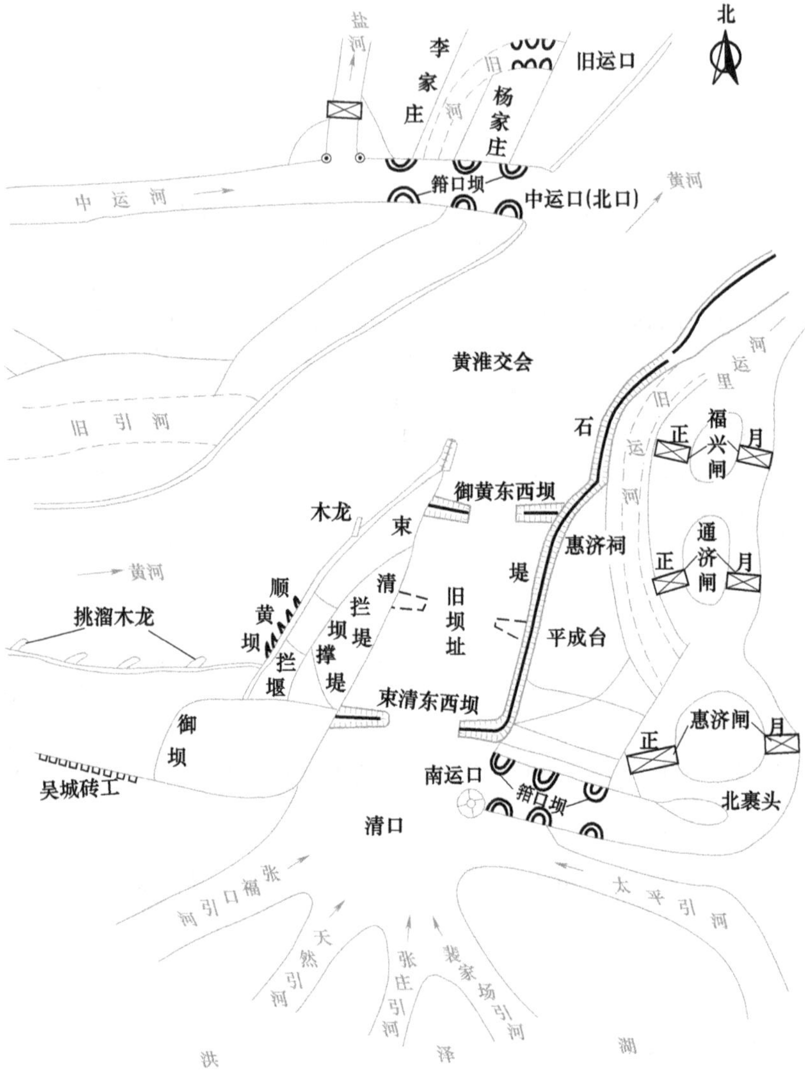

Figure 6.11: Schematic Diagram of Multi-purpose Projects at Qingkou (AD 1785)

the general plan of river improvement in terms of storing clear water to flush out sediments.

Since the reach downstream of Qingkou shared by the Yellow River and the Huaihe River was silted up, the water level in the Huaihe River had to be raised to discharge its clear water to the east.

Figure 6.12: The Hongze Lake Levee

In the 1590s, Pan Jixun came up with the general idea of 'storing clear water to flush out sediments' and considered that raising the water level at Qingkou in the Huaihe River would be beneficial in flushing out the sediment in the riverbed downstream of it, lowering the downstream riverbed elevation, discharging water in the Huaihe River, and ensuring smooth navigation along the canal. As a result, the Gaojia Weir, a large-scale project at that time, was constructed, and the Hongze Lake (see Figures 6.12 and 6.13), an artificial reservoir, was formed due to the weir. In order to ensure its safety, several overflow facilities were constructed on the Gaojia Weir. When floods in the Hongze Lake posed a threat to the safety of the Gaojia Weir, the overflow facilities were used to discharge water in the Hongze Lake into such lakes as the Baoying Lake and the Gaoyou Lake (see Figure 6.14). After a continuous construction in the late Ming Dynasty and the early Qing Dynasty, the Gaojia Weir was gradually perfected until it basically had the same features as modern dams. The Gaojia Weir was a huge masonry dam with a length of approximately

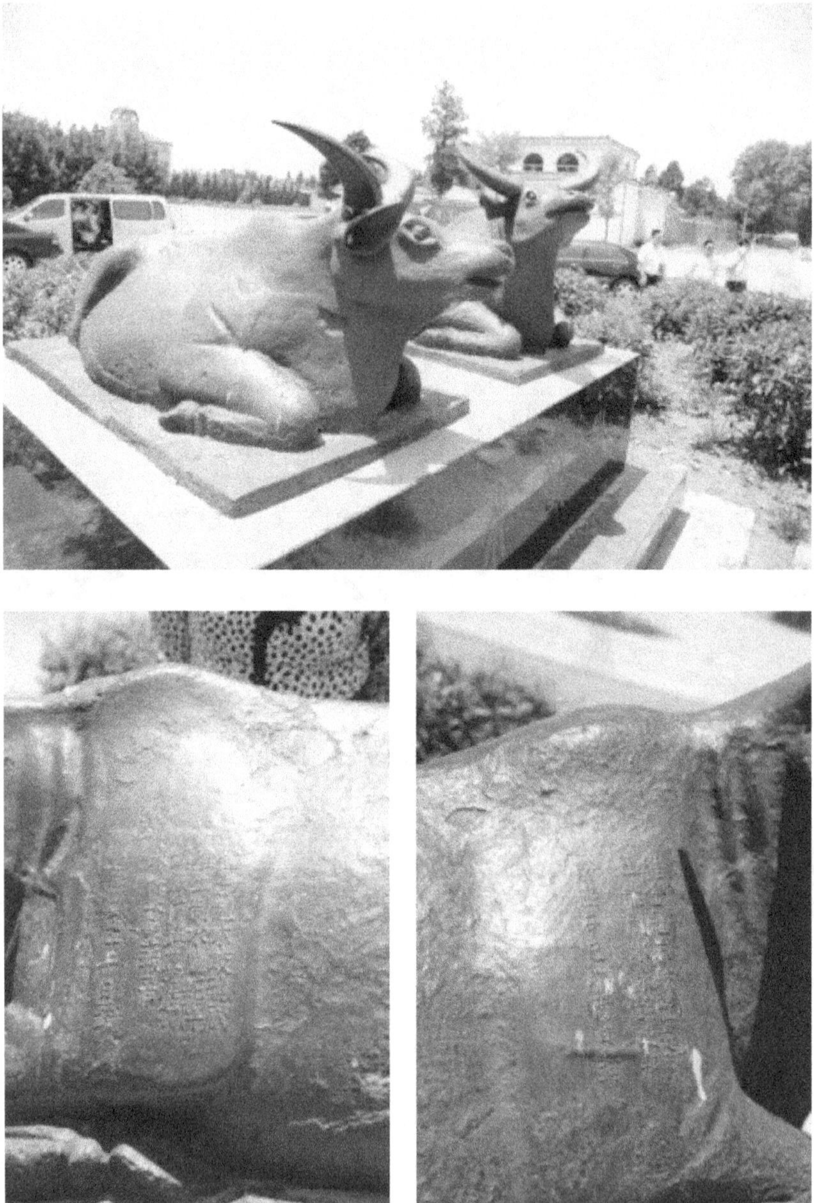

Figure 6.13: Iron Oxen and Inscription on the Hongze Lake Levee

Note: The oxen were made in the 40th year of the Kangxi Era (AD 1701) and was placed on the dike of Hongze Lake.

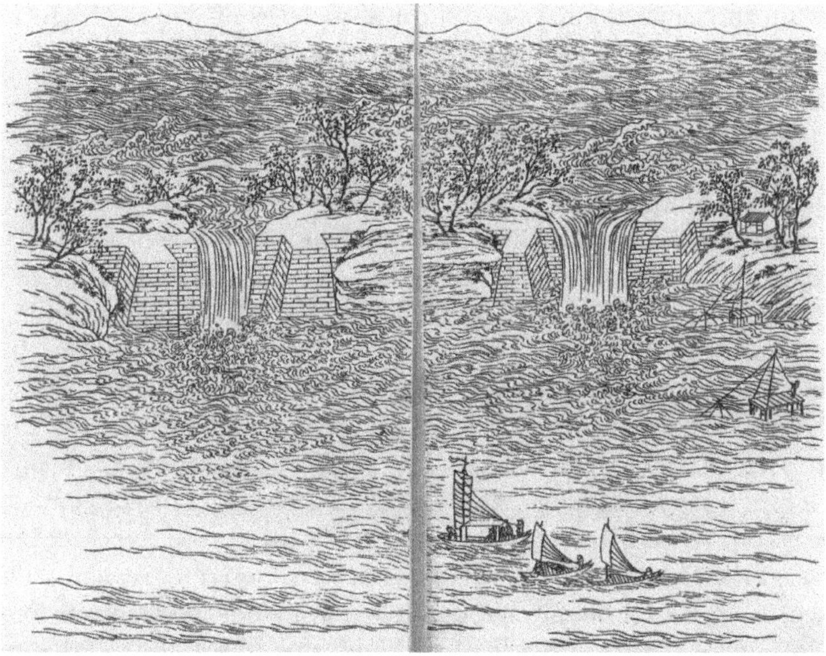

Figure 6.14: Discharge from the Zhi and Xin Weirs on the Hongze Lake Levee

Figure 6.15: The *Status Quo* of Multi-purpose Projects at Qingkou: Waterway Interchange in Huai'an (2013)

60 km and a maximum weir crest elevation of 19 m, and it formed an artificial reservoir with a storage capacity of 3.5 billion m^3, which is notable in the world history of civil engineering.

Because the Yellow River was strong and the Huaihe River was weak, the reach shared by the two rivers, which was downstream of Qingkou, continued to be silted up. In the Qing Dynasty, the problem that water flowed backward to Qingkou had already become very serious. The plan of avoiding using water from the Yellow River and diverting water from the Huaihe River was formulated; structural measures were taken to prevent the sediment-laden water in the Yellow River from flowing backward to Qingkou and the Huaiyang Canal, and clear water in the Huaihe River was diverted to the Huaiyang Canal to aid navigation. In the Qing Dynasty, the Yuhuang Weir and the Shuqing Weir were constructed downstream of Qingkou. When the water level in the Yellow River rose and exceeded that in the Hongze Lake, the two weirs were closed to prevent sediment-laden water in the Yellow River from flowing backward to Qingkou. When the water level in the Huaihe River rose and that in the Yellow River fell, the weirs were opened to discharge floodwater into the Huaihe River. In the Jiaqing Era (the early 19th century), it became an ordinary state of affairs for sediment-laden water in the Yellow River to flow backwards to Qingkou. For navigation, however, the two weirs had to be opened, which further exacerbated the sedimentation at Qingkou. As a result, the plan of 'storing clear water to flush out sediments' ended with failure.

Since the Yellow River bed was silted up, the canal slope around Qingkou increased. In order to ensure boats sailed across the Yellow River smoothly, two types of measures were taken. One was that the northern canal mouth was not directly connected to the Yellow River. Instead, boats went into the Huaihe River via Dawan and then into the Yellow River via Qingkou. This extended the length of the canal and decreased its slope, while at the same time preventing sediment in the Yellow River from entering the canal. The other measure was the construction of multiple sluices and temporary weirs and dams to form a multi-step ship lock system so that boats transporting tribute rice could travel along the canal smoothly.

6.3.3.2 *The Southern Canal Mouths*

The southern canal mouths included the canal mouth at Guazhou and the canal mouth at Yizheng, which were used to connect the canal to the Yangtze River. There was a significant waterhead here, and thus, from the time of the Eastern Jin Dynasty, some control projects had been constructed.

The canal mouth at Yizheng was the entrance of the Yiyang Canal into the Yangtze River (see Figures 6.16 and 6.17). It was an important point that the boats from Huguang, Jiangxi, and the upper Yangtze River area travelled through to transport tribute rice or other materials. The Ouyang Weir constructed in the Eastern Jin Dynasty — namely the canal mouth at Yizheng — is the oldest control project on the canal. In AD 1026, the Jiangkou Weir was reconstructed into a double wooden sluice and again into three masonry sluices 2 years

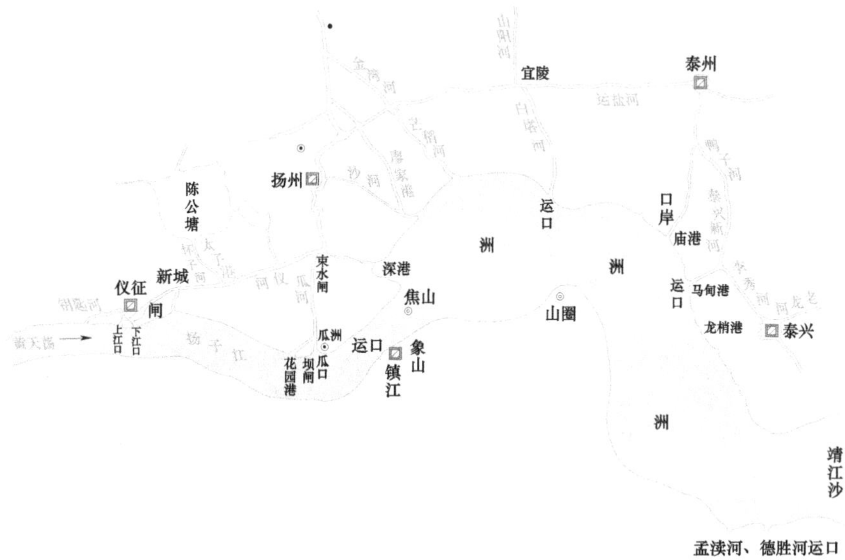

Figure 6.16: Schematic Diagram of Waterways Connected to the Yangtze River Near Yizheng in the Early Qing Dynasty

Note: This quote is from the *Whole Chronology of Huaihe River*, the 79th historical map of Huaihe River.

Figure 6.17: The Current Yizheng Canal Mouth (2013)

Note: In 1972, a large-scale sluice was constructed at the Yizheng canal mouth. The sluice is a multi-purpose project with such functions as diverting water from the Yangtze River for irrigation, discharging floods, and navigation. Nowadays, the sluice remains an important gateway that connects the canal and the Yangtze River.

later. It was typical of the then-multi-sluice projects and abandoned afterwards. In AD 1371, Weirs were constructed on the site of the abandoned Jiangkou Weir that boats transporting tribute rice passed through. In AD 1383, five weirs were constructed in the area south of the Jiangkou Weir and close to the Yangtze River, two of which were used by official boats and bamboo or wooden boats, and the remaining three were used for the transportation of rice and civilian boats. In 1474, Guo Sheng, an official at the Ministry of Works, constructed four sluices connected to the Yangtze River and named them 'Lihekou', 'Xiangshui', 'Tongji', and 'Luosiqiao' between Dongguan in the Li River and Luosiqiao. These sluices were closed when the tide was coming in and opened when the tide was going out. They were abandoned afterwards due to inappropriate management, but then reconstructed as weirs for navigation. In AD 1488, it was decided that these projects would be used both as sluices and weirs. In summer and autumn, when the tide was coming in, they were opened to store tidewater for navigation, while in winter and spring (the dry season), they were closed to hold water from the upper reaches of the canal, and boats were towed over the weirs. In the same year, the Dongguan Sluice and the Luosi Sluice were repaired. In the winter of AD 1499, the Lanchao Sluice was constructed 0.5 km west of Luosiqiao to stop tidewater and divert it to the canal. The Lanchao Sluice was approximately 670 m from the Yangtze River. The Lanchao Sluice was combined with the Luosi Sluice, the Tongji Sluice, and the Dongguan Sluice so that a four-sluice system was formed. When the water level in the lake rose, water was discharged through the four-sluice system. Afterwards, due to leakage, the system was no longer used. Later, these sluices were abandoned and repaired over and over again. In the 1590s, blocked by sedimentation, boats transporting tribute rice were forced to change their route to pass through Guazhou. In AD 1691, the Xiangshui Sluice in Yizheng Canal, the Tongji Sluice, the Luosi Sluice, and the Lanchao Sluice were repaired. After AD 1775, the Yizheng Canal became increasingly shallow, all of the boats transporting tribute rice went through Guazhou, and the Yizheng Canal was only used by boats transporting salt in Huainan (see Figure 6.16).

The Guazhou canal mouth was excavated in the Tang Dynasty. The main route for the transportation of tribute rice from Zhejiang across the Yangtze River was east of the Yizheng canal mouth. In AD 738, when the Yilou River was excavated, the Yilou Weir, namely the Guazhou canal mouth, was constructed. At that time, people had to pay taxes for their boats to pass through the Guazhou Weir, so the annual revenue was extremely high. In the Northern Song Dynasty, when boats were towed over the Guazhou Weir, as many as 22 cows pulled the windlass.

At the beginning of the Ming Dynasty, the weirs constructed in the Song Dynasty could no longer function. In AD 1370, in order to recover navigation, eight weirs were constructed in the Eastern Port and seven weirs were constructed in the Western Port. Later, these weirs were abandoned and repaired many times. From AD 1465 to 1487, the canal was separated into three branches with a shape like the Chinese character '瓜'. The branch in the middle was separated by levees and not connected to the Yangtze River. The branch in the east was called the Eastern Port and the one in the west was called the Western Port, and both of them were connected to the Yangtze River.

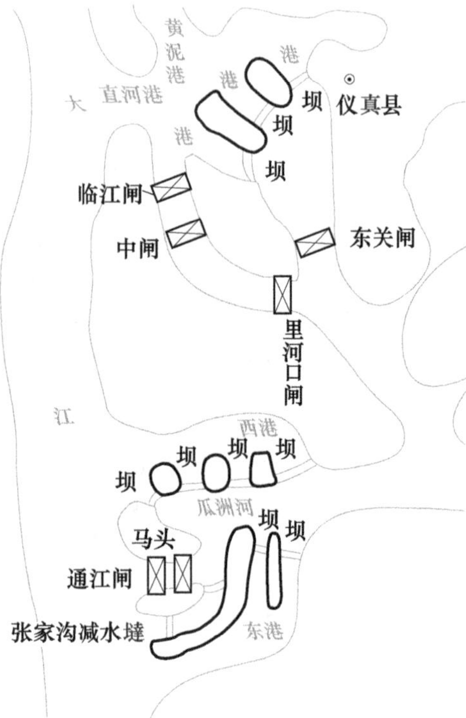

Figure 6.18: Schematic Diagram of the Guazhou and Yizheng Canal Mouth from 1488 to 1505 in the Ming Dynasty

The Eastern and Western Weirs diverted water from the Yangtze River to the canal, and 11 weirs were constructed as control projects (see Figure 6.18). The control projects constructed at the Guazhou canal mouth were weirs, so it was relatively easy for the empty tribute-rice transportation boats coming back from the north of the Yangtze River to be towed over the weirs. However, not only did it take time and manpower to tow the fully loaded boats coming from the south of the Yangtze River over the weirs, but the boats also had to wait in strong winds and huge waves in the Yangtze River for the tide to come in. As a result, in AD 1572, a channel was excavated from Shijiazhou to Huayuangang with a length of 3 km. Meanwhile, two sluices were constructed at Guazhou to be used for boats to enter the Yangtze River. The upstream one was the Guanghui Sluice and the downstream one was the Tonghui Sluice. Afterwards, boats transporting tribute rice could avoid being towed over the weirs.

In AD 1573, it was specified that the sluices should only be open for three months. After all the tribute-rice boats had passed, the sluices were closed and other boats were still towed over the weirs. The following year, the upstream sluice (the Guanghui Sluice) was damaged by insects, so the upstream sluice was changed into the downstream sluice and a middle sluice was constructed at Dunjiazhou. The following year, the Yangziqiao Sluice was constructed.

At the beginning of the Qing Dynasty, the Guanghui and Tonghui Sluices still existed between the Sancha River and Jiangkou, which had a distance of 13.5 km. In AD 1715, the Yangtze River changed its course to the north. The Tonghui Sluice collapsed and the Guanghui Sluice was blocked. As a result, boats transporting tribute rice had to make a detour around the city. In AD 1728, the channel the boats used to make a detour around the city was closed and the boats sailed via the Guanghui Sluice. In 1746, the Guanghui Sluice was closed, and boats sailed along the old Yue Canal at Qinglianan, upstream of the Guanghui Sluice. Perfected in the past dynasties, the Guazhou canal mouth, where one can see Zhenjiang City on the opposite side of the Yangtze River, became a multi-purpose project with such functions as navigation (see Figure 6.19), irrigation, flood discharge, drainage, and protection against the tide. Nowadays, it still functions normally.

6.3.4 *Flood-Discharging Projects on the Huaiyang Canal*

The Huaiyang Canal was a continuous waterway formed through excavating some channels that connected lakes between the Huaihe River and the Yangtze River. It was called the 'Lake Channel' due to the numerous lakes along it. In the Song Dynasty, in order to solve the problem of boats capsizing due to strong winds and big waves while sailing in lakes, levees were constructed between the canal and the lakes along it. After the levee system was formed, rivers flowing from the west to the east were blocked by the levees. As a result, a kind of water-discharging facility called 'Shida' emerged. Shida was used to regulate water.

Figure 6.19: The Current Guazhou Ship Lock (2013)

In the Southern Song Dynasty, the Yellow River changed its course to the south and claimed the lower Huaihe River. As a result, water in the lower Huaihe River was forced to flow south into the Huaiyang Canal. The water flowing from the Huaihe River into the Huaiyang Canal not only made lakes west of the Huaiyang Canal flow east, but also brought the risk of floods to the Lixiahe region for hundreds of years. Therefore, in the Ming and Qing dynasties, five weirs diverting water to the sea and 10 projects diverting water to the Yangtze River were constructed to discharge the floodwater on the Huaiyang Canal.

6.3.4.1 *Shida in the Song Dynasty*

As the levee system was formed in the Huaiyang Canal, Shida emerged, a kind of lateral overflow weir constructed on the eastern levee of the canal. Shida was not large and its construction was simple. It was used to regulate water levels so as to keep enough depth for navigation along the canal, while at the same time avoiding the risk of water flowing over the levees when there was too much in the canal.

In AD 1025, on the northern levee of the canal in Gaoyou, Zhang Lun used a large stone to construct Shida for the lateral overflow, making it potentially the oldest Shida on the Huaiyang Canal. In AD 1035, Wu Zunlu, the Vice Director in charge of transportation in Huainan, 'constructed nineteen sluices named doumen' in places such as Zhenzhou, Yangzhou, Chuzhou, Sizhou, and Gaoyou. The function of doumen is similar to that of Shida. From AD 1118 to 1119, there were 17 document sluices on the banks of the canal in Zhenzhou, Yangzhou, Chuzhou, Sizhou, and Gaoyou. At that time, the construction of water-discharging structures along the Huaiyang Canal reached a climax. From AD 1165 to 1173, Chen Min, a high-ranking military officer, constructed 12 Shidas in Baoying and Gaoyou according to the old system. More than 20 years later, in AD 1194, Chen Sunzhi, a high-ranking official in Huaidong, constructed the long levee between Jiangdu and Huaiying, with 13 Shidas and seven doumens built on the levee.

6.3.4.2 *Ports on the Western Levee of the Canal*

In AD 1680, in order to discharge water in the lakes west of the canal, Jin Fu, an official in charge of rivers, excavated 22 ports on the western levee of the Huaiyang Canal from the west of Gaoyou to Lujin, so as to connect the canal and the lakes west of the canal. Among the 22 ports, five were made of stone and 17 were made of earth. The total length of the 22 ports was 994 m. Normally, the ports were closed by straw and opened to discharge water from the lakes to the canal when the water level rose in the lakes. In AD 1833, the five stone ports were blocked and the 17 earth ports were sometimes blocked and sometimes open. In AD 1921, there were only nine ports connected to the lakes and the total length of the nine ports was 1367 m. Thirteen years later, all the ports were choked up. The following year, a small ship lock was constructed at Yuehegang so as to solve the problem of navigation between the canal and the lakes after the canal and lakes were separated. In the 1930s, nine ports were excavated again at the western bank of the canal, and the total length of these nine ports was 692 m. In 1936, the government planned to close all the ports, but afterwards only two ports were closed due to inadequate supplies of food.

After 1949, as the irrigation channel in Subei had been constructed, the Huaiyang Canal could receive water directly through the Yunnan Sluice. In addition, the five weirs that discharged water to the sea were completely abandoned, which made it possible to separate the Huaiyang Canal from the Gaobao Lake. In 1952, all of the remaining five ports were closed.

6.3.4.3 *Five Weirs Discharging Water to the Sea and 10 Projects on the Yangtze River*

In AD 1847, the Qingkou exit of the Huaihe River was basically abandoned due to sedimentation. In AD 1851, the levee of the Hongze Lake broke at Liba (modern-day Sanhekou). Since then, the Huaihe River has completely changed its course. The levee at Liba had never been repaired. The Huaihe River went through the Hongze Lake, came out of it at Liba, and then flowed into the Baima Lake, entered the Lixiahe region via the canal, or went down via lakes around Gaoyou and finally flowed into the Yangtze River at Sanjiangkou, Jiangdu, after meeting the canal at Shaobo. After the Huaihe River changed its course, the flood-control pressure was increased. The main facilities to regulate floods were the five weirs discharging water from the canal to the sea and 10 projects on the Yangtze River.

The construction of the five weirs to discharge floods began in AD 1679 and was basically completed in AD 1757 (see Figure 6.20). The

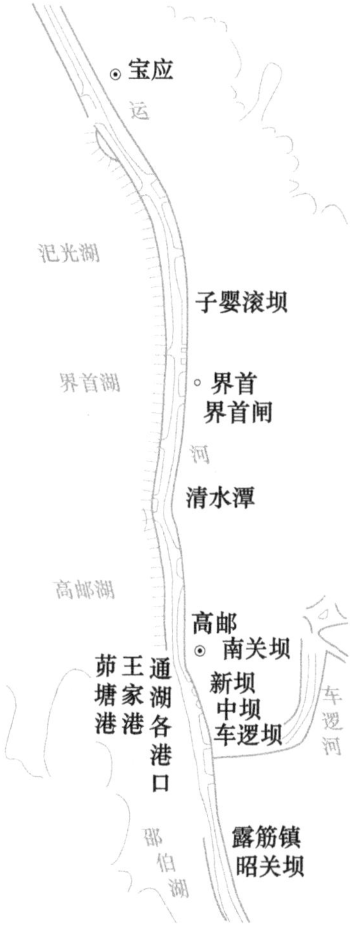

Figure 6.20: Schematic Diagram of the Five Weirs During the Reign of Emperor Qianlong of the Qing Dynasty

five dams were named Nanguan, Nanguanxin, Wulizhong, Cheluo, and Shaoguan. The operating rules for the five weirs were formulated based on how much water was discharged from the Hongze Lake. For instance, if increasing amounts of water was discharged through the Wugun Weir on the levee of the Hongze Lake and the overflow water depth of the Cheluo Weir and the Nanguan Weir reached 1.17 m, the Wuli Weir was opened; if the overflow water depth of the Cheluo Weir and the Nanguan Weir continued to increase and reached 1. 67 m, the Nanguanxin Weir was opened. The Cheluo and Nanguan Weirs were always open to discharge floods at any time. In order to ensure safe navigation, the remaining three weirs were not opened unless the water depth in the canal was above 1.67 m.

The operating rules for the five weirs were later revised many times. In AD 1828, the operating rules stated that if the water depth in the canal reached 4.27 m, the Cheluo Weir was opened; if it reached 4.4 m, the Nanguan Weir was opened; if it reached 4.53 m, the Wulizhong Weir opened; and if it reached 4.67 m, the Nanguanxin Weir opened.

In AD 1855, people such as Li Hongzhang, the Governor of the Liangjiang region, took Liqiu, the beginning of autumn, as a dividing point. Before Liqiu, if the gauge reached 4.47 m, the Cheluo Weir was opened; then, for each additional 0.13 m, the Nanguan Weir, the Wulizhong Weir, and the Nanguanxin Weir were opened in turn. After Liqiu, if the gauge reached 4.27 m, the same protocol was followed as during Liqiu.

In order to control the stage and flood-discharging volume, straw and earth were put on top of the weirs. When the weirs were opened, the straw and earth would be washed away and the masonry weirs below the straw and earth were not affected. The five weirs played an important role in discharging floods in the Huaihe River, especially after the middle of the Qing Dynasty and after the Huaihe River changed its course and flowed south. The water discharged by the five weirs, however, brought serious disasters to the Lixiahe Region.

After the Huaihe River changed its course in AD 1851, the sluices and weirs in the Huaiyang Canal that diverted water from the canal to the Yangtze River were the main facilities of discharging floods in the Huaihe River. The projects that discharged floods to the Yangtze River were gradually constructed as well, and all of them were completed during the Daoguang Era in the Qing Dynasty. The 10 projects (see Figure 6.21) that discharged floods to the Yangtze River were named the Jinwanbei Sluice, the Jinwangun Weir, the Dongxiwan Weir, the Fenghuangqiao Sluice, the Wantou Sluice, the Liaojiagougun Weir, the Shiyanggougun Weir, the Dongjiagougun Weir, the Mangdao Sluice, and the Chujia Weir. The name of each project was changed during different periods.

Figure 6.21: Schematic Diagram of the 10 Projects During the Reign of Emperor Qianlong of the Qing Dynasty

Downstream of the projects, there were channels to divert the water discharged from the Huaihe River to the Yangtze River. The operating rules of the projects were formulated based on gauge. Downstream of the 10 projects, there were six channels to divert water to the Yangtze River, and each of the channels went from the north to the south. From Wantou to Xiannyumiao, an ancient canal ran from the west to the east, receiving water from the six channels on the northern bank and discharging water to four channels on the southern bank. The four channels were named the Liaojia Channel, the Shiyang Channel, the Dongjia Channel, and the Mangdao Channel. The four channels continued to go down, and three of them flowed into the Mangdao Channel at Shilidian and Hanjiadu. As a result, the four channels became one channel with the collective name of the 'Mangdao Channel'. Two channels, namely the Mangdao Channel and the Shiyanggou Channel, flowed into the Shatoujia River and then met at Sanjiangying, finally becoming one channel. In short, the number of the channels that discharged water to the Yangtze River was four at first, then two, and finally one. The channels were wide and shallow in their upper reaches and then became narrow and deep in their lower reaches. Their water flow was slow in the upper reaches and then became fast in the lower reaches, and their riverbed elevation was high in the upper reaches and low in the lower reaches. The scouring force of the channels has not changed for hundreds of years. Nowadays, the channels are still functioning normally.

Chapter 7
The Jiangnan Canal

7.1 Introduction

The Jiangnan Canal ('Jiangnan' literally means 'south of the Yangtze River' in Chinese) was one of the first reaches of the Grand Canal to be constructed, and it has had the longest continuous operation time and the best natural conditions. Its excavation started during the Spring and Autumn Period (770–476 BC), its entire channel was realigned during the Sui Dynasty, and its southern reach became gradually straightened after the Yuan Dynasty. The region that the Jiangnan Canal runs through is abundant in rainfall, with a dense river network and relatively rich water resources. The history of the Jiangnan Canal reflects the whole process of the construction and operation of the Grand Canal in China. As one of the areas with the most developed economy in China, the region where the Jiangnan Canal is located boasts important cities and towns, all of which are closely related to the canal and embellish it like bright pearls.

7.2 Basic Conditions

7.2.1 *Overview*

The Jiangnan Canal starts from Zhenjiang City of Jiangsu Province in the north and ends in Hangzhou City of Zhejiang Province in the south with a total length of 340 km. It connects the Yangtze River, Lake Taihu, and the Qiantang River. The whole channel of the Jiangnan Canal remains navigable today. Building on the ancient canal, the Jiangnan Canal has formed a crisscross navigation network that enables ships to go from north to south and from west to east, and it continues to play an important role in transportation across the region.

The main channel of the Jiangnan Canal was excavated through full use of the topographical conditions and water resources, with riverbed elevation ranging from 0 to 2 m and the steepest longitudinal slope of only 0.05 per cent between Zhenjiang City and Wuxi City. Lake Taihu was the water source for two thirds of the canal. At the northern and southern ends of the canal, the requirement for navigation could basically be met by means of water-source projects. With the changes of the natural environment in the region, the southern reach of the canal gradually became straight after the Yuan Dynasty. Multi-sluice projects used widely in the Song Dynasty were typical of the technical level of the Jiangnan Canal, which once represented the most advanced level of navigation in China — and even in the world. The multi-sluice projects could concentrate where the topographical elevation differences were greatest along the canal, and they controlled the water level by operating together. This enabled ships to pass steadily though the sluices with a small waterhead between two adjacent ones, while at the same time refilling their chambers or receiving water from their chamber by 'water-regulating ponds'. Thus, they could reduce water loss when ships passed through them. Compared with a single sluice, the multi-sluice projects helped ships pass through them horizontally, which could lower risks while at the same time reducing water loss due to the opening of the sluices. Compared with weirs and dams, the multi-sluice projects made it possible for ships to pass through them without surmounting weirs or dams, saving time and reducing cost.

The Jiangnan area has been one of the regions with the most developed economy in China since ancient times. Since the Tang and Song dynasties, it had been not only a place from where tribute rice was transported to the feudal dynasties of past ages, but also a region with advanced levels of handicraft and commerce. Since the 10th century, the Jiangnan area has been the centre of the country's economy, providing the external conditions necessary for the longstanding Jiangnan Canal. Important cities such as Zhenjiang, Wuxi, Suzhou, Hangzhou, and Jiaxing are located along the Jiangnan Canal, whose development has promoted the development of those cities. As an important part of the Grand Canal system, the Jiangnan Canal has provided a channel for the exchange of goods with external areas, while at the same time creating conditions for the exchange and fusion of the regional economy and culture. All the cities along the Jiangnan Canal are relatively large and prosperous. The cities and canal are inextricably linked; the urban water supply and drainage are achieved by means of the canal, which strengthens the relationship between the cities and the canal. In addition to cities, a large number of villages and towns have been set up along the canal. The human social environment and natural ecological environment have formed the landscape of the Jiangnan region with distinct regional characteristics, which have become the important cultural feature of the region ever since the Ming Dynasty. The Jiangnan Canal has not only witnessed the history of the economic and social development of the Jiangnan region, but also helped shape that history by assisting the formation of the cultural landscape of the towns and villages with distinct regional features.

The benefits gained by the navigation of the Jiangnan Canal were significant. The amount of tribute rice was approximately 378,000 tons/year in the Ming Dynasty, more than one third (126,000 tons) of which was from the Jiangnan region and Zhejiang. After 1472, this proportion increased to nearly 50 per cent (approximately 170,000 tons). In the Ming Dynasty, the Jiangnan Canal had to take on the task of transporting 100,000-odd tons of tribute rice. In the Qing Dynasty, the same system was adopted and the annual amount of tribute rice was approximately 378,000 tons, some 170,000 tons of which was from the Jiangnan region and some 59,000 tons from Zhejiang. In the Qing Dynasty, the Jiangnan Canal had to take on the task of transporting approximately 229,000 tons of tribute rice.

After the system of tribute-rice transportation was abolished at the end of the Qing Dynasty, especially after the preliminary development of modern industry, the Jiangnan Canal continued to play an important role in transportation. According to the statistics of the Canal Committee of the Republic of China, from 1908 to 1917, the annual average number of the ships that sailed along the Hangzhou reach of the canal was 13,446, with the transportation amount reaching 536,432 tons. Meanwhile, the transportation amount of the railway between Shanghai and Hangzhou was only 950,528 tons in 1931.

The Jiangnan Canal has withstood the evolution of the natural environment and river systems. Before the Tang Dynasty, the surface area of Lake Taihu was larger than it is now, and the canal and Lake Taihu were linked together. The present Tanglu (a kind of levee) and Ditang (a kind of canal), which are several kilometres southeast of Lake Taihu, are actually the former banks of the lake.

During the Tang and Song dynasties, the reach of the canal between Suzhou and Pingwang was separated from Lake Taihu through the construction of the Wujiang Levee. Afterwards, navigation in the canal was no longer affected by winds and waves or by the rising and falling of the water level during the flood and dry seasons in Lake Taihu. To some extent, the levee impeded the discharge of floodwater, although many culverts were built under it. Before the Ming Dynasty, floods in Lake Taihu were discharged into the sea through the Wusong River. In addition to the construction of the Wujiang Levee, the sedimentation in the lower reach of the Wusong River made it more difficult to discharge the floods in the Lake Taihu into the sea, so after the Song Dynasty, the main water issues in the Jiangnan region were flooding and waterlogging due to the poor discharge of the floods in Lake Taihu. It was not until the Ming Dynasty that the Taipu River was excavated to replace the Wusong River for the purpose of discharging floods in Lake Taihu. Then the issue of discharging floods in Lake Taihu was solved.

Before the Yuan Dynasty, the Xiatang River (with the Qiantang River as its water source) was one of the main channels of the Jiangnan Canal. During the Yuan Dynasty, because the northern bank of the Qiantang River was silted up and it became difficult for the Qiantang River to provide water for the Xiatang River, the Xiatang River (with the Dongtiaoxi River as its water source) became one of the main channels of the Jiangnan Canal. The evolutionary history of the Jiangnan Canal is an important point of reference for understanding the evolution of the natural environment in the region.

7.2.2 *Natural Background*

Linked with the Yangtze River in the north and the Qiantang River in the south, the Jiangnan Canal runs through the Lake Taihu basin in a north-to-south direction.

Topographically, the elevation of the land is higher in the west than in the east, while it is higher in the coastal area than in the middle of the Taihu Plain. As a result, the area is a saucer-shaped basin with the Lake Taihu at its centre. In the north, south, and east, the Lake Taihu basin is a waterfront close to rivers and the sea with a flat topography. The land elevation in the north, however, is different from that in the south. The northern area of the basin is a hilly area/plain with a land surface elevation of 10–50 m in the hilly areas and 6–10 m on the plain. The southern and eastern parts of the basin are a low-lying area with a land surface elevation of 2–5 m, where a lot of rivers and lakes are located and waterlogging disasters occur frequently. The western part of the basin is a mountainous area with a land surface elevation of approximately 200–500 m. The Jiangnan Canal runs through the northern plain and the low-lying area of the basin.

The Lake Taihu basin is located in the south of the northern subtropical zone and the north of the central subtropical zone. It has an annual average rainfall of between 1,000–1,200 mm and is rich in water resources. Affected by the circulating current of the monsoon, its rainfall has an uneven spatial and temporal distribution. With Lake Taihu as its centre of water regulation and storage, the river systems of the Jiangnan region receive water from the western mountainous area. The Jingxi River of the west route and the Tiaoxi River of the south route are the main river systems of the Jiangnan region. The eastern river systems serve as the discharging channel of Lake Taihu and originally had three main river systems, which were historically called the 'Three Rivers' — the Dong River, the Lou River, and the Wusong River. Around the 11th century, the Dong River and the Lou River were choked with silt. As a result, the Wusong River has become the main channel that discharges water from Lake Taihu into the sea. The southern part of the Jiangnan Canal horizontally crosses the flood-discharging channels of Lake Taihu, and therefore it has relatively rich water source, while the northern part or the reach between Zhenjiang and Changzhou is the ridge of Lake Taihu and the Yangtze River, and therefore has few level-crossing river systems, which makes water supply to the Jiangnan Canal a major problem.

7.2.3 *The Historical Evolution of the Jiangnan Canal*

7.2.3.1 *From the Spring and Autumn Period to the Qin Dynasty: The Connection of Natural Rivers*

There is no clear-cut record of when the Jiangnan Canal began to be constructed. There is a dense river network in the Lake Taihu basin, and natural rivers in it are navigable (see Figure 7.1). Back in the late Spring and Autumn Period, Wu and Yue, two states in the Zhou Dynasty, started excavating channels between natural rivers, which formed the preliminary canal. According to records in the historical literature, from the Spring and Autumn Period to the Qin Dynasty,

Figure 7.1: The Canal in Wuxi City — the Old Taibodu Canal

Notes: The Taibodu Canal is located in Wuxi City, Jiangsu Province, connecting the Jiangnan Canal and the Li Lake. According to legend, it was excavated in the late Shang Dynasty (the 12th century BC) and is considered the country's oldest canal project by some researchers (see *The History of the 3,000-Year Canal in China* by Ji Guohuang, Encyclopaedia of China Publishing House, 2008). It is now an urban river in Wuxi City.

the excavated canals in the Jiangnan region included the Baichidu and Wugu channels in the north of Suzhou city that were connected to the Yangtze River and roughly parallel to the present canal, the Yuzhaodu channel around Tongjiang in Changzhou City and Wuxi that was connected to Lake Taihu, and the Ling channel that started in the south of Suzhou City, ran through Jiaxing City, and ended in Hangzhou City. At the beginning of the Qin Dynasty, the Yangtze River, the Qiantang River, and Lake Taihu were already connected by artificial channels.

7.2.3.2 *From the Western Han Dynasty to the Sui Dynasty: Preliminary Formation of Canals*

In the Han Dynasty, an artificial channel was excavated between Suzhou City and Jiangxing City, but its exact route cannot be verified. During the Period of the Three Kingdoms (AD 220–265), a ripe south-to-north navigational channel already existed, which began in the southeast of the Lake Taihu basin and flowed into the Yangtze River in Zhenjiang City with a route roughly parallel to the present Jiangnan Canal.

In AD 245, Sun Quan ordered Chen Xun to mobilise 30,000 people for the excavation of Pogangdu, a canal from Jurong to Yunyang (modern-day Danyang), which replaced the navigational channel named 'Jingkou' and enabled ships to reach Nanjing directly, avoiding the risk of sailing on the Yangtze River. Pogangdu was a summit canal on which 12 weirs were constructed to regulate water flow and balance altitude differences. The weirs are the oldest ones recorded in literature.

During the Wei–Jin Southern and Northern dynasties, great efforts were made to improve the navigational channel in Changzhen and construct the Dingmao Weir at the canal mouth and the Benniu Weir in Changzhou and excavate an artificial pond — Lian Lake — in Danyang. As a result, a navigational system composed of tidal gates and diversion sluices, water-regulating weirs, and the artificial pond was initially established for the northern part of the Jiangnan Canal. During this period, many weirs and dams were constructed to regulate flow in the reach connected to the Qiantang River, and technologies for towing boats over weirs and dams were relatively ripe. According to the records in a book entitled *The History of the Jin Dynasty*, in AD 362, buffalo towed boats over weirs around the Qiantang River; the fees charged for this service went to the government, which shows the amount of water in the canal as well as the scale, management extent, and economic benefits of it.

In AD 610, Emperor Suiyang systematically improved the Jiangnan Canal on the basis of the existing canal, and the route of the Jiangnan Canal was determined; the route has not changed significantly since. The water source and regulation works in the canal continued to be improved. At that time, the Jiangnan Canal began at the Jingkou canal mouth in the Yangtze River, ran south through Qu'a (modern-day Danyang, Jiangsu Province), Piling (modern-day Changzhou, Jiangsu Province), Wuxi, Wujun (modern-day Suzhou, Jiangsu Province) and Jiaxing, finally reaching Yuhang (modern-day Hangzhou, Zhejiang Province) via the Shangtang River.

7.2.3.3 *In the Tang and Song Dynasties: The Formation of an Independent Canal System and Development of Navigational Engineering Technology*

China's economy and politics had reached an unprecedented height in the Tang and Song dynasties. The demand for tribute-rice transportation increased, resulting in a higher demand for the control of water levels and supply in the canal. With the construction of the Wujiang Levee, the Jiangnan Canal became increasingly independent. The application of multi-sluice projects (similar to the present engineering system of ship locks) demonstrated that the project construction and management level of the canal had reached a new historical record. The controlling projects were constructed in the Tang Dynasty, such as the Jingkoudai Weir in Zhenjiang, the Chengtingdai Weir in Danyang, the Wangting Weir in Wuxi, and the Chang'an Weir in Chongde. In the Song Dynasty, the number of controlling projects increased and most of the projects were reconstructed as multi-sluice projects.

In the Tang and Song dynasties, great attention was paid to the construction and maintenance of the water-source projects on the Jiangnan Canal, and the army was often mobilised to dredge the channel that led to the Yangtze River. Water was mostly diverted from Jingkou of the Yangtze River to the northern part of the Jiangnan Canal, and the Lian Lake was the main storage pond. In AD 813 (the 8th year of the reign of Yuanhe in the Tang Dynasty), the Mengdu Channel in Changzhou, which led to the Yangtze River, was excavated so as to divert water from the Yangtze River to the canal for the navigation. The water source of the Shangtang River in the southern part of the Jiangnan Canal was mainly dependent on the Qiantang River and the West Lake. In the Song Dynasty, the water source became inadequate because of issues such as the sedimentation in the West Lake and the impact of the sediment-laden tidal water in the Qiantang River. Thus, new water sources had to be found. In AD 1086, Su Shi, a Prefectural Governor at the time, dredged the West Lake and the Yanqiao River and constructed a sluice in the Mao Shan River to divert water to the Shangtang River. In AD 1247, after a serious drought, the West Lake dried up and the Shangtang River ran dry. As a result, one more channel was excavated to divert water from the Dongtiaoxi River to the Tang River in Yuhang so as to store the water in the West Lake. Because of the unstable water source of the Shangtang River, a channel between Nanfengkou of Deqing County and Bei-xingqiao was later excavated to divert water from the Dongtiaoxi River to the Xiatang River.

(A) *Construction of the Wujiang Levee: The Jiangnan Canal Becomes a Truly Independent Engineering System*

Before the 7th century, the area around Wujiang, in the eastern part of Lake Taihu and the southern part of Suzhou, was submerged in water. There was no boundary between the canal, the Wusong River, and Lake Taihu. The boats for tribute-rice transportation sailed along the bounds of Lake Taihu. The area was boundless and indistinct in the flood season, but it was full of shoals in the low-flow season, which made navigation difficult. In AD 810, Wang Zhongshu, a Prefectural Governor in Suzhou, started constructing a long levee with aggregated rock between Wujiang and Suzhou and along the eastern bank of the Lake Taihu. In 1042, Li Yuqing, an official under the county magistrates who administered lawsuits in Suzhou, continued to construct a long levee of 40 km between the Wujiang River and Pingwang; since then, the canal and Lake Taihu have been separated. The long levee between Suzhou and Pingwang was called 'Wujiang Levee' or 'Wujiang Bank' by later generations. The Wujiang Levee impeded the discharge of floodwater in Lake Taihu into the Wusong River, and therefore culverts and bridges were constructed under it to discharge floodwater. The top of the levee served as the road for boat trackers. This levee became 'a multi-purpose project of Wujiang' with comprehensive benefits, such as navigation, flood control, and transportation. The completion of the levee indicated that the Jiangnan Canal had become a truly independent artificial system.

(B) *The Management of the Jiangnan Canal and Application of Multi-Sluice Projects in the Song Dynasty: The Zenith of Engineering Technology and Management on the Grand Canal*

In the Tang Dynasty, weirs served as the main control projects on the Jiangnan Canal. In the Song Dynasty, a series of control gates or multi-sluice projects were used to replace weirs in the Jiangnan Canal. With the advent of the multi-sluice projects, the transportation organisation of a fleet forming into columns and the operation management of projects appeared. The application of the multi-sluice projects represented the highest level of the navigational engineering technology at that time. Similar projects appeared in Europe over 600 years later. In AD 1072, Chengxun, a Japanese Buddhist monk, came to China for worship. He took a boat to go up, starting in Hangzhou and sailing through the Jiangnan Canal. According to his records, the Chang'an Gate in Yanguan County, the Shanqing Gate in Jiaxing, the Wangting Gate in Wuxi, the

Benniu Gate in Changzhou, and the Jingkou Gate in Zhenjiang were all multi-sluice projects.

7.2.3.4 *The Evolution of the Canal in the Yuan, Ming, and Qing Dynasties*

(A) *Partial Change in the Route of the Jiangnan Canal in the Yuan Dynasty*

Since the Song Dynasty, the southern part of the Jiangnan Canal had had two branches south of the Chongde (modern-day Chongfu Town, Tongxiang City). The northern branch was the Xiatang River that turned west at Chongde, ran through Tangqi, turned south and finally reached Hangzhou. The southern branch — the main branch — was the Shangtang River, which ran south through the Chang'an Gate, turned west, and finally reached Hangzhou. The water source of the Shangtang River relied on the West Lake and the tidewater from the Qiantang River. In the Southern Song Dynasty, the northern bank of the Qiantang River was broken many times by strong tides, and water-shortage problems in the Shangtang River arose frequently. Thus, in AD 1247, the Xiatang River was dredged; after this, it became increasingly important. In the Yuan Dynasty, due to the rising of the sand beach in the northern bank of the Qiantang River, it became difficult to divert water from the Qiantang River into the Shangtang River. Thus, the Shangtang River was filled with sediment. During the reign of Zhi Zheng, the Shangtang River was abandoned because it became unfavourable for navigation, and thus efforts were made to improve the Xiatang River. At the end of the Yuan Dynasty, a new channel starting at Tangqi was excavated to extend the Xiatang River towards the south, to the Jiangzhang Bridge in Hangzhou. The new channel was used to replace the Shangtang River as the main route for the transportation of tribute rice.

(B) *The Jiangnan Canal in the Ming and Qing Dynasties*

During the Ming and Qing dynasties, the central government was responsible for the management of the Jiangnan Canal. At the very

Figure 7.2: Schematic Diagram of the Zhenjiang Canal and its Branches Leading to the Yangtze River

Notes: The canal between Zhenjiang and Changzhou passes over the ridge between the Yangtze River and the Taihu Lake, which is the reach with the highest elevation and suffers the most serious problem in terms of water supply. In the Ming Dynasty, since the main route was not navigable, the branches became the main channels for the transportation of tribute rice, including:

1. Mengdu, an excavated canal that, in the east, was connected to the main route at modern-day Benniu Town, Changzhou City. In the north, it led to the Yangtze River and was called the 'Baita River' after passing through the Yangtze River. Mengdu was originally excavated by Mengjian, a Prefectural Governor in the 8th year of the reign of the Yuan He Emperor in the Qing Dynasty. The initial intended functions were the irrigation by diversion and water supply for navigation with an irrigation area of 4, 000 ha. In the Ming Dynasty, it became the main channel for the transportation of tribute rice and a control sluice was constructed where the Mengdu Canal was connected to the main route in the south.
2. Desheng New Canal, which used to be known as 'Lietang Canal'. It was located west of Changzhou and opposite the North New Canal after passing through the Yangtze River. It was excavated in the Southern Song Dynasty and a sluice in it was constructed near the Yangtze River. The sluice was abandoned in the Yuan Dynasty and reconstructed in 1370. In 1391, the canal was dredged and renamed as the 'Desheng New Canal'. Its original function was irrigation by diversion, but after the Xuan De Era, the canal began to be used for the transportation of tribute rice.

beginning of this period, the central government designated officials to manage the canal. In the Wanli Era, it was managed by the Director General of the canal under the premier's leadership, and then by the Commissioner of the canal. In the Qing Dynasty, only the canal mouth in Zhenjiang was managed by the Director General of the Jiangnan Canal. Except for that, other channels and projects were under the jurisdiction of local governments.

In the Ming and Qing dynasties, emphasis was put on the improvement of the northern part of the Jiangnan Canal. The Changzhen Canal was often filled with silt to the extent that it was not navigable. As a result, several branches, such as the Mengdu Canal, the Desheng New Canal, and the Jiangyin Canal, were dredged to divert water or serve as alternative channels for navigation (see Figure 7.2), while at the same time greatly improving the capacity of Lian Lake as a storage pond. Of these canals, the Mengdu Canal (see Figure 7.3) was used to transport tribute rice for the longest time. It was not until AD 1578 that the branches — including the Mengdu Canal — were gradually abandoned because the Danyang Canal became navigable after being dredged.

Until the Qing Dynasty, the over-100-km canal between Jingkou and Wuxi was often filled with silt because of the tidewater from the Yangtze River. The government then decided to undertake small-scale dredging work every year and large-scale dredging work every 6 years. In the late Qing Dynasty, the Jingkou reach was silted up so seriously that boats could not sail in it even when there were weak

Figure 7.2 (***Continued***): (Figure on facing page)

3. The Jiangyin Canal, which was formed by dredging a natural river. Between Wuxi and Jiangyin, there were a number of harbours, such as the Huangtian Harbour and the Xia Harbour. As early as the Tang dynasty, a sluice was set up for diversion at the Huangtian Harbour. The sluice was expanded and reconstructed many times in the Song and Yuan dynasties. In the Ming Dynasty, it was one of the routes for the transportation of tribute rice. It was not until the Wanli Era that the Zhenjiang Canal was recovered. The boats going up with heavy goods passed through Jingkou in Zhenjiang, while the boats going down without any goods navigated the Jiangyin Canal. Boats also navigated the Jiangyin Canal while the Jingkou Canal was being extensively dredged.

Figure 7.3: Location Where the Mengdu Canal is Connected to the Jiangnan Canal (2009)

Notes: After the Ming Dynasty, the Mengdu Canal gradually lost its function as a route for the transportation of tribute rice and was gradually silted up. Its present name is the 'Xinmeng Canal' and it has a width of 20 m and a normal water depth of less than 1 m.

tides in the Yangtze River. The Lian Lake also was gradually silted up because of constant reclamation, losing its function of supplying water for navigation. In the 9th year of the Daoguang Era, the Lian Lake was improved to supply water for navigation, but it made little difference. The branches in the northern part of the Jiangnan Canal were also silted up and thus had to be dredged many times. After Daoguang, because Qingkou was not navigable, tribute rice was gradually transported through navigating the sea. The tribute rice was transported from Shanghai to Tianjin via the sea.

After the 5th year of the Xianfeng Era, the canal in Shandong Province was broken by the Yellow River and the Grand Canal was only an empty title. At that time, the Jiangnan Canal experienced the transition from being the economic lifeline of the country to a regional canal. Because of the lack of management of the canal, the reach between Jingkou and Dantu dried up in the Tongzhi Era. The reach between Dantu and Wuxi was only approximately 30 cm deep. The Lian Lake was silted up and thus abandoned. During the Guangxu Era, the canal was dredged and improved, but as the system of tribute-rice transportation was abolished, the Jiangnan Canal never recovered to its past level.

7.3 Key Projects

The Jiangnan Canal is 340 km long, connecting the Yangtze River, the Lake Taihu, and the Qiantang River. During the Spring and Autumn Period, natural river systems were connected locally. A continuous channel was formed during the Qin Dynasty, connecting the Yangtze River and the Qiantang River. During the Sui Dynasty, through a unified channel improvement, the Jiangnan Canal became a stable route for the transportation of tribute rice. During the Song Dynasty, an independent and perfected project system was formed. Afterwards, the Jiangnan Canal was used for the transportation of tribute rice for hundreds of years. The regulation and control of water projects ensured the continuous operation of the Jiangnan Canal.

7.3.1 *Waterway Engineering*

The Jiangnan Canal is located in an area with a dense river network, and the natural rivers have been historically navigable (see Figure 7.4). The canal was formed through artificial excavation and river training and became the main waterway for the transportation of tribute rice in different historical periods. The Jiangnan Canal consisted of a main route and many branches. The rivers in the cities that the Jiangnan Canal ran through were also an integral part of it.

7.3.1.1 *The Main Route of the Jiangnan Canal*

The main route of the Jiangnan Canal is from Zhenjiang (see Figure 7.5) to Hangzhou, with a total length of 340 km. It connects the Yangtze River, the Qiantang River, and the water system of Lake

Figure 7.4: The Canal Near Hangzhou (The Xiatang River, 2007)

Note: The Jiangnan Canal is still a busy goods transportation route, mainly handling construction materials.

Taihu. The Jiangnan Canal has a higher elevation at its two ends and a lower elevation in the middle. The planning of the main route gave full consideration to the topographical conditions, enabling a stable transition of the elevations of the land surface along the route. The average longitudinal slope of the canal was 0.016 per cent between Zhenjiang and Suzhou, while it was 0.006 per cent between Suzhou and Hangzhou. The maximum average longitudinal slope of the canal was 0.05 per cent between Zhenjiang and Wuxi, where the densest concentrations of control projects were located. The main route basically connected the main cities in the Jiangnan region, playing an important role in promoting and supporting the development and prosperity of these cities. With the social development and progress of engineering technology, the reach south of Lake Taihu had a tendency towards gradual cut-off, which was especially significant in the Qing Dynasty.

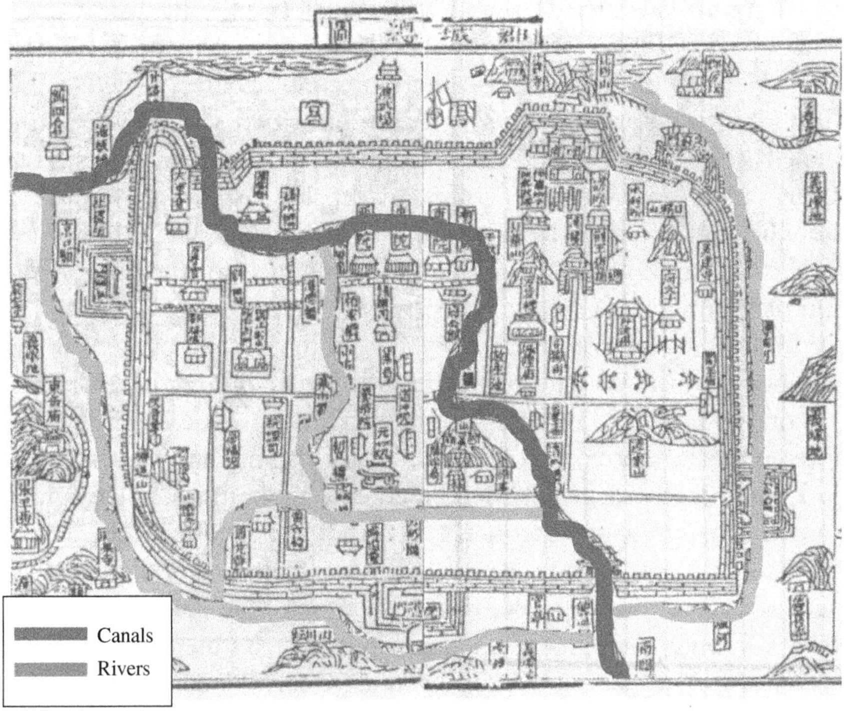

Figure 7.5: The River System of Zhenjiang City During the Reign of the Qianlong Emperor in the Qing Dynasty

Notes: Zhenjiang City is located south of the Yangtze River. The Jiangnan Canal runs through the city, intersecting with the rivers of the city. The canal runs out of the city to its west and then intersects with the Yangtze River at Jingkou.

7.3.1.2 *Multi-Purpose Branches*

The Jiangnan region is characterised by a dense river network. In the Jiangnan region, in addition to the north–south main route of the canal, there are many branches that are either natural rivers or excavated regional canals with multiple functions, such as navigation, water supply for the main route, and irrigation. The branches are an integral part of the Jiangnan Canal.

The Zhenjiang Canal had the highest elevation with an inadequate water source. As a result, a number of branches leading to the Yangtze River were excavated to divert water from the Yangtze River to the Zhenjiang Canal. The branches were also navigable, including the Mengdu Canal, the Desheng New Canal, and the Jiangyin Canal. The canal around Youming between Changzhou and Zhenjiang had been in a state of disrepair, and therefore the branches served as the main channels for the transportation of tribute rice. Where the branches intersected horizontally with the main route or the Yangtze River, sluices were set up to control water flow. After the Wanli Era, the Zhenjiang Canal was recovered, and therefore some branches, such as Mengdu, no longer served as the channels for the transportation of tribute rice.

In the Qing Dynasty, the water source of the reach around Danyang became extremely inadequate, and the weakening of Lian Lake's ability to store water due to reclamation exacerbated the issue of inadequate water in the northern parts of the main route that had the highest elevation. The branches, such as Mendu, could only form a supplementary water source for the reach around Changzhou. Therefore, in Zhenjiang, channels such as the Dantu Canal, the Yue Canal, and the Jiuqu Canal were excavated or dredged to divert water from the Yangtze River to the main route of the Jiangnan Canal. Some canals used to serve as a waterway for navigation, but they were later abandoned or boats went up other channels. The original canals became the branches of the Jiangnan Canal, such as the Pogangdu Canal, the Shangrongdu Canal, the Rouge Canal (see Figure 7.6), and the Shangtang River. The Pogangdu Canal was located in Jurong County, which was excavated in AD 245 to connect the Danyang Canal and the QinHuai River and served as a shortcut to avoid the danger of the boats sailing along the Yangtze River while transporting the tribute rice to Nanjing. In the Nan Liang Era, the Shangrongdu Canal was excavated north of the Pogangdu to replace the Pogangdu Canal. In the Chen Era, the Shangrongdu Canal was abandoned and the Pogangdu Canal was recovered. It was not until China was unified and Kang City (modern-day Nanjing) was constructed in the Sui Dynasty that the Pogangdu Canal and the Shangrongdu Canal were both abandoned. At the beginning of the Ming Dynasty, Nanjing became the capital of the country and the Rouge Canal (also named the 'Tianshengqiao Canal') was excavated in Lishui County west of Lake Taihu to connect the QinHuai River and the Xuxi Canal. As a result, boats could enter Lake Taihu,

Figure 7.6: The Rouge Canal (2010)

Notes: The Rouge Canal, a summit canal, is 15 km long. It was excavated in AD 1393. At the beginning of the Ming Dynasty, Nanjing was the country's capital. In order to avoid the danger of tribute-rice boats sailing against the current in the Yangtze River, the Xuxi Canal was improved and the Rouge Canal was excavated, making it possible to navigate the canal from the western part of Lake Taihu to Nanjing via the Xuxi Canal, the Shijiu Lake, and the Qinhuai River. The Rouge Canal passed over the ridge between the Qinhuai River basin and the Shijiu Lake with a maximum depth of over 20 m. It was very difficult to excavate the Rouge Canal. The reach upstream of Shigang was excavated by burning stones and watering, and finally the Rouge Canal had a width of approximately 15 m and a depth of up to 20 m. In AD 1421, the capital was moved to Beijing and the Rouge was no longer used for the transportation of tribute rice. The Rouge Canal was used for less than 30 years.

leave the Xuxi Canal, and reach Naning via the Rouge Canal and the QinHuai River. After the capital was moved to Beijing in the Yongle Era, the Rouge Canal was not used any more.

Some channels, such as the Fengkou Canal, that had been excavated to solve the water-source problem of a certain reach of the Jiangnan Canal also simultaneously aided navigation. Some of the branches ran perpendicular to the main route of the Jiangnan Canal, serving as the projects for regional flood control or waterlogging control while at the same time aiding navigation. Some branches are local canals connected to the main route, such as the Ditang Canal (see Figure 7.7) and the Zhihetang Canal. Most of the branches have multiple functions, such as water supply and navigation, but different

Figure 7.7: The Ditang Canal in Nanxun (2010)

Notes: The Ditang Canal was located at the southern margin of Lake Taihu, connecting the Tiaoxi River and the main route of the Jiangnan Canal. The Ditang Canal was excavated by Yinkang, a Prefecture Chief in the Western Jin Dynasty. It passed through 72 small rivers. In AD 792, it was reconstructed.

branches and even the same branch at different times put emphasis on different functions.

The main route between Suzhou and Hangzhou goes south to Jiaxing City and then turns west. In the Qing Dynasty, in order to shorten the distance and cut down on the cost of transportation, the reach of the Xiatang River between Lanxitang and Baimatang was cut off so that some boats did not need to make a detour via Jiaxing, going northeast directly to Pingwang via Shimen. This cut-off waterway did not replace the position of the Xiatang River that ran through Jiaxing and served as the main channel for navigation.

Many branches were important water-source projects for the Jiangnan Canal while simultaneously serving as the alternative waterways for the reaches of the main route. The branches greatly raised the reliability of navigation on the Jiangnan Canal; at the same time, as an extension of the Jiangnan Canal, it enlarged the scope of navigation along the river network. In addition, the branches had the function of supplying water for farmland irrigation around them.

These multi-purpose branches embody the comprehensiveness of the planning of the canals and the flexibility of the project's operation and management. In other words, the maximum benefit is gained by the minimum facilities.

7.3.2 *The Lian Lake as a Storage Pond*

The Changzhou–Zhenjiang reach of the Jiangnan Canal was located on the boundary between Lake Taihu and the Yangtze River, and it had a serious water-source problem. The branches leading to the Yangtze River, such as the Mengdu Canal, could supply water to the reaches of the Jiangnan Canal in Changzhou, where the elevation was relatively low, while the Lian Lake (see Figure 7.8) served as the water-source project for the reach of the Jiangnan Canal in Danyang, where the elevation was highest. Historically, every time the Lian Lake had good management and maintenance and was functioning normally, the northern reach of the Jiangnan Canal tended to be navigable. On the contrary, when the Lian Lake was not managed well or could not function to its fullest extent, the Jiangnan Canal was not

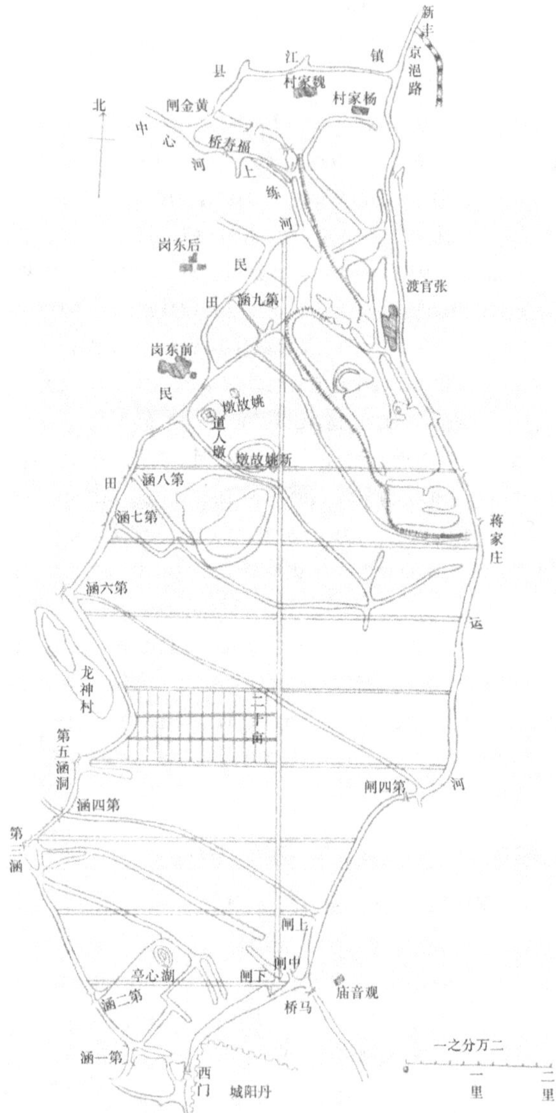

Figure 7.8: The Lian Lake and its Relationship with the Jiangnan Canal

Notes: This map is from a paper entitled 'Investigation on the Lian Lake in Danyang and Planning of Proposed Model Reclamation Area' in the journal *The Lake Taihu Water Resources Quarterly* (Vol. 4), published in 1931. It was drawn by the Jiangnan Water Resources Bureau based on an on-site survey. Most of the lake area had become farmland then due to reclamation; it was known as the Lian Lake Farmland, but the outline of the Lian Lake and its relationship with the Jiangnan Canal are the same as they were in the historical period.

navigable. The Lian Lake was constructed in the Western Jin Dynasty and was originally named the 'Lian Pool'. It was located north of Danyang County and its original purpose was to store water for irrigation. In the Tang Dynasty, the perimeter of the Lian Lake was 20 km and its irrigation capacity reached hundreds of hectares.

In the Southern Song Dynasty, the Lian Lake was separated into the upper lake and the lower lake. Three sluices were set up in the upper, middle, and lower parts of the Lian Lake to ensure water sources for the Lian Lake, connecting the upper and lower lakes and supplying water for navigation on the main route of the Jiangnan Canal, respectively. At the beginning of the Yuan Dynasty, the Lian Lake was entirely reclaimed for farming. In AD 1294, the Lian Lake was excavated again, and thereafter it was dredged many times. At the beginning of the Ming Dynasty, there were three lateral weirs (called 'doumen') in the Lian Lake, two of which were in the upper Lian Lake and one of which was in the lower Lian Lake. In the Jianwen Era, all of the three weirs were abandoned. The Saidouhu Weir, a lateral weir, was then used to divert water to the lower Lian Lake. In the lower Lian Lake, the upper doumen, the middle doumen, and the lower doumen were constructed around the site connected to the canal. Afterwards, a levee was constructed in the southwestern part of the Lian Lake. At the end of the Ming Dynasty, four sluices were set up in the Lian Lake. In the Daoguang Era in the Qing Dynasty, the Lian Lake was used to supply water for the Beidantu Canal. In AD 1868, the four sluices in the Lian Lake (see Figure 7.9) were repaired. In 1883, they were repaired again. After the end of the 19th century, the Lian Lake was gradually abandoned due to its poor management and a decrease in the quantity of tribute rice being transported.

As a storage pond, the Lian Lake did not have favourable topographical conditions. Sheng Fusheng, a scholar in the Qing Dynasty, summarised the Lian Lake as follows:

> Generally, lakes have their respective stable water sources, but the Lian Lake does not. Water in the Lian Lake is from the overland flow of the Long Mountain, which is not controllable. Mostly, lakes are navigable, but the Lian Lake is not. Unlike the Tao Lake and the Ge Lake, in which there is always water, the Lian Lake has a relatively high terrain,

Figure 7.9: The Site of the Cement Sluice with Five Gates in the Lian Lake (1990s)

Notes: In the 1940s, a cement sluice with five gates was reconstructed at the location of the third sluice of the Lian Lake. The cement sluice has now been abandoned. The Lian Lake no longer exists, the area having become a residential area and farmland.

> is shallow, and easily dries up. Usually, lakes are the product of a natural process, but the Lian Lake is an artificial lake. Water of the Dachen Creek in the Long Mountain flows into the upper Lian Lake and the lower Lian Lake, and then is blocked by levees and stored in the Lian Lake. Water level in the Lian Lake is regulated by sluices and culverts that have been constructed along the levees. How much water is discharged from the Lian Lake depends on the time of inflow into the lake and water use provided by the lake. By contrast, however, lakes with deep valleys and big pools downstream of them can freely discharge water from them into the downstream areas.

The Lian Lake was often unable to help in diverting water to the Jiangnan Canal due to the reclamation of the area for farming. The Lian Lake was abandoned because the local tyrannical landlords or peasants diverted water from it to irrigate the nearby farmland, or because of cultivation in the lake area. In 1928, the Lian Lake was turned into farmland. At that time, there was a complicated channel network for irrigation and drainage and water was discharged into the main route of the Jiangnan Canal when waterlogging occurred.

7.3.3 *Control Projects*

The elevation of the Jiangnan Canal was higher at its two ends and lower in the middle. There was an especially big difference in elevation in the Changzhou-to-Zhenjiang, there was a big elevation difference. In addition, inadequate water supply was a serious problem. As a result, it was necessary to construct some projects to control the water depth and discharge. According to the historical literature, the oldest control project is a weir named Dingmao, which was constructed at Jingkou in the Jin Yuan Era. In the Northern Song Dynasty, there were nine control projects recorded along the main route of the Jiangnan Canal (see Figure 7.10). This is the period with the largest number of control projects taking place. Afterwards, some dams and weirs were reconstructed as gates or multi-sluice projects, and some were abandoned in the process of integrating the project systems. In terms of the type of projects, the development of the control projects in the Jiangnan Canal experienced three stages — weir or dam, sluice, and then multi-sluice. These represent the three aspects of the navigation engineering technology in the ancient canals and will be introduced in detail in the following sections. Since the beginning of the 20th century, the Jiangnan Canal has been entirely improved and widened, the level of navigation has been raised, and most of the historical hydraulic structures — such as ancient weirs, dams, and sluices — have been demolished.

7.3.3.1 *Technology Involved in Dam Construction and Pulling Boats Over Dams*

Weirs or dams were the control projects on the Jiangnan Canal that were operated for the longest time. 1,800 years ago, 12 weirs were set up along the Pogangdu Canal, and they were operated until the end of the Qing Dynasty. The structure of the weirs or dams were made of masonry. They were dam rivers. Sailing boats or ships need to mount a weir or dam to reach the other side.

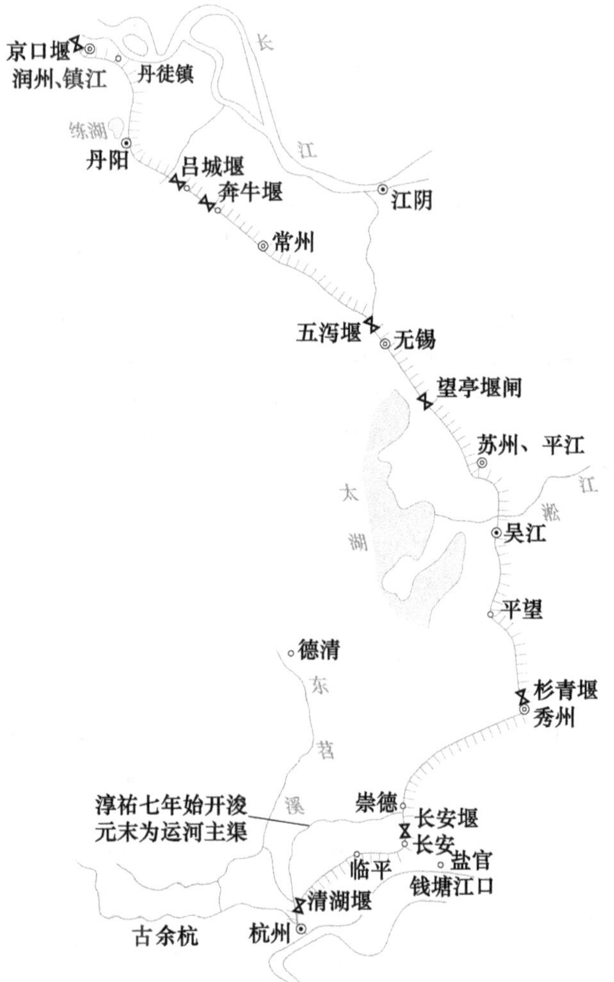

Figure 7.10: Schematic Diagram of the Control Projects in the Jiangnan Canal in the Northern Song Dynasty (see *History of the Beijing-to-Hangzhou Grand Canal* by Yao Hanyuan)

Notes: During the Northern Song Dynasty, there were nine control projects recorded in literature along the main route of the Jiangnan Canal. They were the Jingkou Weir in Zhenjiang, the Maiting Weir and the Lyucheng Weir in Danyang, the Benniu Weir in Changzhou, the Wangting Weir in Wuxi (now within the boundaries of Suzhou), the Shanqing Weir in Jiaxing, the Changan Weir in Chongde, and the Hanjiang Weir and Qinghe Weir in Hangzhou. This was the period when there was the largest number of control projects.

It was not until the 4th century that the technology of using animals to pull boats over a weir or dam was applied to the Jiangnan Canal. According to the records in the book entitled *The History of the Jin Dynasty*, in the 1st year of the Longhe Era (AD 362), in Haiyan (present-day Haiyan City) and Qiantang (now the eastern part of Hangzhou), buffalo pulled boats over dams in the canal. The owners of the boats had to pay service fees, which indicates that this method of pulling boats over dams was already well-established. In AD 488, Emperor Qiwu planned to raise the fees for pulling boats over dams. The original revenue per dam or weir per year was 1 million Qian (the currency at the time). Some people suggested that the fee was doubled or more, so the annual revenue from four weirs or dams would be nearly 10 million Qian. This indicates the boom of navigation in the canal at that time.

In AD 1072, a Japanese monk named Chengxun came to China for worship. He took a boat to go up from Hangzhou along the Grand Canal. He was surprised at the way boats were pulled over dams. According to the records in his diary, when boats arrived at the Wangting Weir, the Benniu Weir, and the Jingkou Weir, buffalo pulled them over. Pulley tackles were set up on both sides of a weir. The pulley tackle on each side rotated due to the pull from seven or nine buffalo. The boats were pulled over the weir by the ropes connected to the pulley tackles. During the reign of the Qianlong Emperor in the Qing Dynasty, a British diplomat visited and was deeply impressed by the canals in Ancient China. He specifically mentioned the weirs and dams on the canal between Suzhou and Hangzhou and described how the boats were pulled over (see Figure 7.11). This method was still in use in the early 20th century (see Figures 7.12 and 7.13).

(A) *Sluice (a single-sluice project)*

In the Tang Dynasty, sluices were constructed to replace weirs/dams as control projects. In the Song Dynasty, multi-sluice projects appeared, but they did not last for long because it was difficult to manage them. Afterwards, multi-sluice projects were replaced by single-sluice projects or even by weirs/dams. In the Ming and Qing

Figure 7.11: A Boat Being Pulled Over a Weir, Drawn by a British Diplomat During the Reign of the Qianlong Emperor in the Qing Dynasty

Figure 7.12: Buffalo Pulling a Boat Over a Dam in the Hangzhou Area at the End of the Qing Dynasty (1900s)

Figure 7.13: A Boat Being Pulled Over a Dam in the Jiangnan Region During the Period of the Republic of China (1920s)

dynasties, almost all of the control projects in the Jiangnan Canal were sluices.

A single-sluice project concentrates water-level difference on its upstream and downstream sides. When the sluice gate is opened, water flows very quickly. As a result, it is necessary to use equipment and tools to control the boats. Single-sluice projects can be dangerous to boats, and they tend to release a lot of water for boats to pass.

During the reign of the Qianlong Emperor in the Qing Dynasty, William Alexander, a painter in the British Diplomatic Corps, portrayed a single-sluice project as follows:

> The water head between the upstream and downstream sides of the sluice is over 6 feet. On the upstream side, water surface is only one feet below the boat's board. The sluice consists of a pair of stone blocks, both sides of which have a slope of approximately 40°. When passing through the sluice, the boat is towed over the sluice to the downstream side by capstans. Usually two capstans are enough, but

> sometimes if a boat is heavy, four or even six capstans are required. In this case, the capstans have to be firmly installed through boring holes on the ground. When a boat passes through the sluice, the rope of a capstan will be linked to an iron hoop in the rear of the boat, which connects the capstan and the boat. The people who push the capstans need to keep adjusting the rope so that the boat can pass through the sluice smoothly. Because of the weight of the boat itself, it will drop into the water on the downstream side with a very fast speed. As a result, to prevent a large amount of water from flowing into the boat, a solid braided fabric is fixed to the front of the boat.

As control projects, the single-sluice projects have some shortcomings, but they are convenient for navigation compared to weirs/dams and easy to operate and manage compared to the multi-sluice projects. As a result, after the Song Dynasty, the single-sluice projects became the main stream control projects on the Jiangnan Canal.

(B) *Multi-Sluice Projects*

In the Song Dynasty, the application of multi-sluice projects indicated that the country's science and technology advancement in navigation engineering had reached its historical summit. In Europe, this kind of project appeared only 600 years later.

The multi-sluice projects in the Jiangnan Canal are complete project systems. A multi-sluice project is usually composed of a group of control sluices, water-regulating pools, and channels connecting the water-regulating pools and the canal. The water-regulating pools are water-source projects. There are usually two water-regulating pools, termed the 'water-supplying pool' and 'water-storing pool'. The water level of the water-supplying pool is usually higher than or the same as the high water level of the chamber (usually the upstream chamber) connected to the water-supplying pool. The water-supplying pool supplements water in the chamber, from which water is released whenever a boat passes it so that the water level in the chamber can be raised and kept the same as on the upstream side, which gets the chamber ready for the next boat.

The water level of the water-storing pool is usually lower than or the same as the low water level of the downstream chamber. The water-storing pool is designed to store water from the chamber when the water level in the chamber is lowered for a boat to pass, which prevents water from being released to the downstream side. If necessary, water in the water-storing pool can be lifted into the water-supplying pool for repeated use. The water source of a water-regulating pool is rainwater from places with a higher elevation, the water lifted from places with a lower elevation (e.g. via the Lyucheng sluice), or tidewater diverted into the water-regulating pools when the tide is coming in (e.g. via the Jingkou sluice). A volume of water at least equal to the volume of a chamber is released by an ordinary multi-sluice project for each boat. The water-regulating pool, however, made it possible to utilise the water released from a chamber repeatedly.

According to the diary of Chengxun, a Japanese monk who came to China for worship in AD 1072, there were two multi-sluice projects in the canal between Hangzhou and Zhenjiang at that time, namely the Chang'an Three-Sluice Project in Yanguan County and the Shanqing Two-Sluice Project in Xiuzhou (modern-day Jiaxing). When the boat he took arrived at the Chang'an Project, the upstream two sluices (in the south) were opened first. When the boat passed through the first chamber upstream and entered the second chamber downstream, the upstream two sluices were closed and the third sluice downstream (in the north) was opened. The waterhead between the upstream and downstream sides of the third sluice was approximately 1.7 m, so the boat did not pass through the third sluice until the water level in the second chamber became equal to that in the downstream side of the third sluice. In this case, the three-sluice project was actually used as a two-sluice project.

Along the Chinese Grand Canal, the Jiangnan Canal is the reach in which there used to be the most multi-sluice projects. According to the literature, the Jingkou Sluice, the Lyucheng Sluice, the Benniu Sluice, the Shanqing Sluice, and the Chang'an Sluice were reconstructed as multi-sluice projects during the Song Dynasty. Of those, the Jingkou Sluice, the Lyucheng Sluice, the Benniu Sluice, and the Chang'an Sluice had water-regulating pools. The completed Jingkou

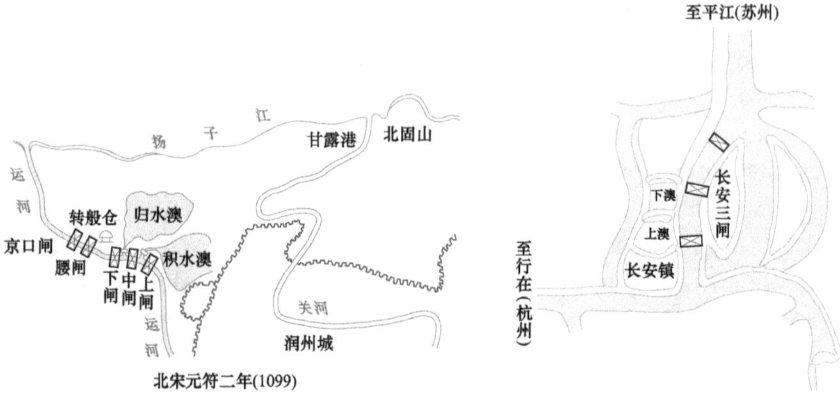

(a) 京口闸 (b) 长安闸

Figure 7.14: Schematic Diagram of the Jingkou and Chang'an Multi-Sluice Projects in the Northern Song Dynasty

Notes: 1. Jingkou, where the Jiangnan Canal was connected to the Yangtze River, was located northwest of Zhenjiang. In a north-to-south order, the Jingkou Multi-Sluice Project in the Jiangnan Canal consisted of the Chao Sluice, the Yao Sluice, and the Lower, Middle, and Upper Sluices. The Chao Sluice was 0.5 km away from the Yangtze River. Between the Chao Sluice and the Yao Sluice was the reach into which tidewater was diverted from the Yangtze River; that had a length of 600 m, and it was where boats waited until the tide was coming in the Yangtze River. The Yao Sluice was approximately 400 m away from the Lower Sluice and was abandoned in the Southern Song Dynasty. The Lower, Middle, and Upper Sluices were located at the ridge of the Yangtze River and Lake Taihu, with an interval of 120 m between them. The two water-regulating pools made full use of the features of the local topography, which were somewhat higher than the Jiangnan Canal but could be reached by the tidewater. As a result, they could store water when the tide was coming in and supply water to the Jiangnan Canal when it was short. The chamber between the Lower Sluice and the Middle Sluice was connected to the water-storing pool, and the one between the Middle Sluice and the Upper Sluice was connected to the water-supplying pool.

2. The Chang'an Multi-Sluice Project was located in the Shangtang River in Jiaxing, from where boats could go west to Hangzhou, north to Jiaxing, and south to the Qiantang River. At the Chang'an Multi-Sluice Project, water level was higher in the south than in the north. In a south-to-north order, the Chang'an Multi-Sluice Project was composed of the Upper, Middle, and Lower Sluices. By taking advantage of the local topography, two water-regulating pools that were connected to two chambers were constructed west of the Jiangnan Canal.

Sluice and Chang'an Sluice are typical of multi-sluice project systems (see Figures 7.14 and 7.15).

Compared with a single-sluice project, a multi-sluice project concentrates the elevation difference at one place and then adjusts the

Figure 7.15: The Location Where the Jiangnan Canal is Connected to the Yangtze River Near the Jingkou Multi-Sluice Project (2010)

elevation difference through several control projects elsewhere. By means of chambers that can adjust its water level to reduce the water-head between the upstream side and the downstream side, boats can pass through sluices smoothly, which has greatly enhanced the safety of navigation. On the other hand, the joint operation of multiple sluices can largely reduce water loss that would otherwise be caused by boats to pass through the sluices.

The advent and operation of multi-sluice projects indicates that the technology in the navigation engineering of canals had developed by leaps and bounds. Meanwhile, the greatest benefit could be gained only by strict operation and management. Under the social conditions and management level in China at the time, however, the multi-sluice projects had been challenged by the political system. High-ranking officials and bigwigs took advantage of the privilege that fresh tributes had priority over other goods, opening sluice gates at will, which disturbed the normal order of navigation and made the water-regulating pool unable to play a due role. After the Song Dynasty, all the multi-sluice projects in the Jiangnan Canal were abandoned and replaced by single-sluice projects.

7.3.4 *The Wujiang Multi-Purpose Project*

The construction of the Wujiang Levee (see Figure 7.16) began in the Tang Dynasty and was completed in the Song Dynasty. Its completion determined the final route of the canal between Suzhou and Wujiang, ended the history of using Lake Taihu for the transportation of tribute rice, and indicated that the Jiangnan Canal had become an independent water project system.

The reason why the Wujiang Levee was called the Wujiang multi-purpose project is that it had comprehensive water benefits.

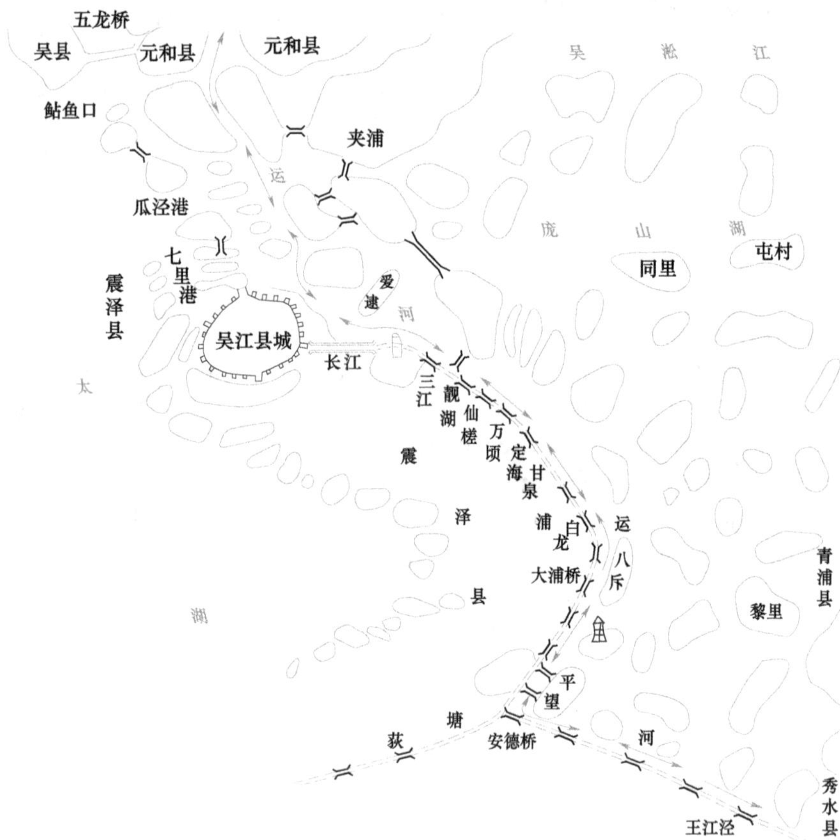

Figure 7.16: Schematic Diagram of the Wujiang Levee and the Jiangnan Canal

Figure 7.17: Baodai Bridge in 2010

According to statistics, along the Wujiang Levee there were 37 bridges and 134 culverts, of which the Baodai Bridge (see Figure 7.17) in Suzhou and the Chuihong Bridge in Wujiang were multi-span bridges with a length of over 300 m. These bridges and culverts, together with the levee, constituted a multi-functional water project system.

The Wujian Levee not only improved the transportation in water and land, but also provided necessary conditions for the large-scale reclamation of low-lying swamps. It played a significant role in reducing the threat of floods to the low-lying cropland east of Lake Taihu and enhanced the capacity of Lake Taihu to stop and store floodwater. The construction of the long levee, however, had an impact on the discharge of floods in Lake Taihu to a certain extent. The area east of the Jiangnan Canal was gradually silted up so that it became a piece of land. The Wusong River was gradually silted up as well, which once resulted in flooding in Lake Taihu and thus posed a huge threat to the Jiangan region. After the Ming and

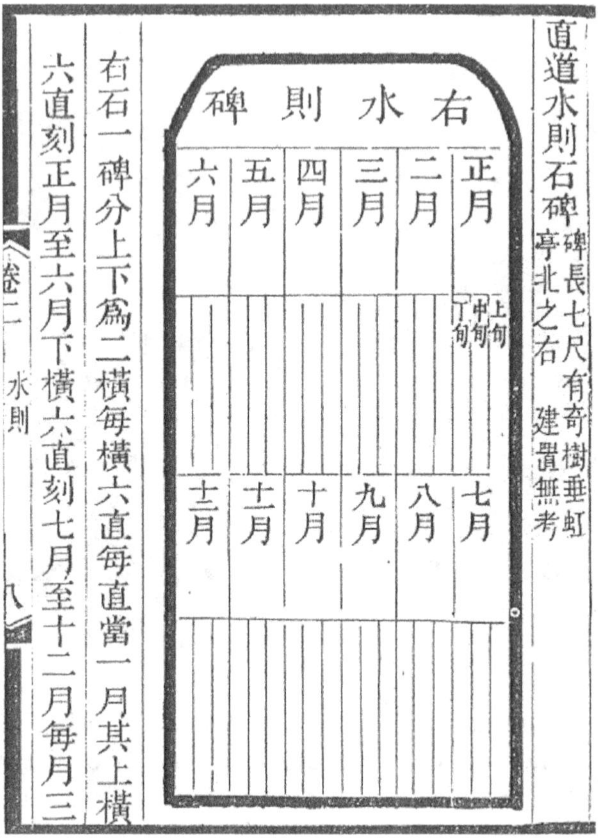
直道水則石碑碑長七尺有奇樹垂虹亭北之右建置無考

右水則碑

正月 二月 三月 四月 五月 六月

上旬 中旬 下旬

七月 八月 九月 十月 十一月 十二月

右石一碑分上下為二橫每橫六直每直當一月其上橫

六直刻正月至六月下橫六直刻七月至十二月每月三

卷二 水則 八

Figure 7.18: Pattern of the Wujiang Stage-Gauging Stela

Qing dynasties, instead of the Wusong River, the Huangpu River became the channel of discharging floods from Lake Taihu. In order to observe and record the water levels in Lake Taihu and the Wusong River, some stage-gauging stations named 'stage-gauging stela' or 'Wujiang staff gauge' (see Figure 7.18) were set up at places such as the Chuihong Bridge. The stage-gauging stelas provided information for the accurate assessment of floods and flood disasters.

Chapter 8
The Zhedong Canal

8.1 Introduction

The Zhedong Canal refers to the southernmost part of the Grand Canal, connecting the rivers around it and the sea. It is not only one of the important starting points of the ancient Silk Road at sea, but also one of the canals that was formed first and operated for the longest continuous period. The construction of the Zhedong Canal began with the excavation of the Shanyin Canal of the Yue Kingdom in the Spring and Autumn Period, but it was not until the Song Dynasty that the Zhedong Canal became a continuous waterway with a perfect system of control projects. During the Southern Song Dynasty, the country's capital was established in the southern city of Lin'an (today's Hangzhou). The movement of the political centre to the south played a key role in promoting the development of the Zhedong Canal. The Mingzhou (modern-day Ningbo City) Port was one of the most important ports along the coast of East China. As a result, the Zhedong Canal became a link between the inland canal and the sea in terms of navigation.

8.2 Basic Conditions

Located in the Xiaoshan–Shaoxing–Ningbo Plain along the south bank of Qiantang Bay, the Zhedong Canal is surrounded by mountains and the sea. Some natural rivers, such as the Puyang River (before the Ming Dynasty) and the Cao'e River in the Xiaoshan–Shaoxing Plain, flow roughly from south to north and finally into the sea. The Yuyao River flows from west to east and finally into the sea.

The Zhedong Canal consisted of artificial channels and natural rivers, including the artificial reach from Xiaoshan to Shangyu and the natural reach from Shangyu to Ningbo, using such natural rivers as the Yao River and Yong River for navigation. Affected by region-specific hydrological conditions such as physical geography, typhoons, and the tide, the control projects of the Zhedong Canal also had significant and unique features. In the process of the formation of the Zhedong Canal, water projects were used to integrate the regional river network and form a canal system with comprehensive water benefits such as navigation, flood control, drainage, water diversion, and irrigation. The canal system played a crucial role in the regional water structure.

The Zhedong Canal is a crucial component of the Grand Canal system. Its development and evolution has consistently followed the development of Chinese society. It has witnessed the development of the economy and society in the country, especially within the Zhedong region.

In the Spring and Autumn Period, the Yue Kingdom was powerful and prosperous, which made it possible for them to develop water transportation with the excavation of the Shanyin Waterway as the starting point. In the Northern and Southern dynasties, China's political and economic centre moved to the south, which raised the position of the Zhedong Canal and made it develop rapidly. As a result, a channelised continuous waterway symbolised by the application of a large number of weir projects was initially formed. In the Tang and Song dynasties, the country was powerful and prosperous and deepened its economic and cultural exchange with foreign countries, which encouraged the Zhedong Canal, a waterway that connected the inland and the sea, to finally become a complete water transportation system. In the Southern Song Dynasty, the country's centre moved to the south again, which laid a solid foundation for the Zhedong Canal as a national project. In the Yuan, Ming, and Qing dynasties, the Zhedong Canal was eventually incorporated into the Chinese Grand Canal system with the establishment of the northern and southern tribute-rice transportation systems.

8.2.1 *Natural Background and General Location of the Zhedong Canal*

Located on the Xiaoshan–Shaoxing Plain of the back land on the south bank of Qiantang Bay, the Zhedong Canal (see Figure 8.1) is in the northeast of Zhejiang Province. It is against the northern foot

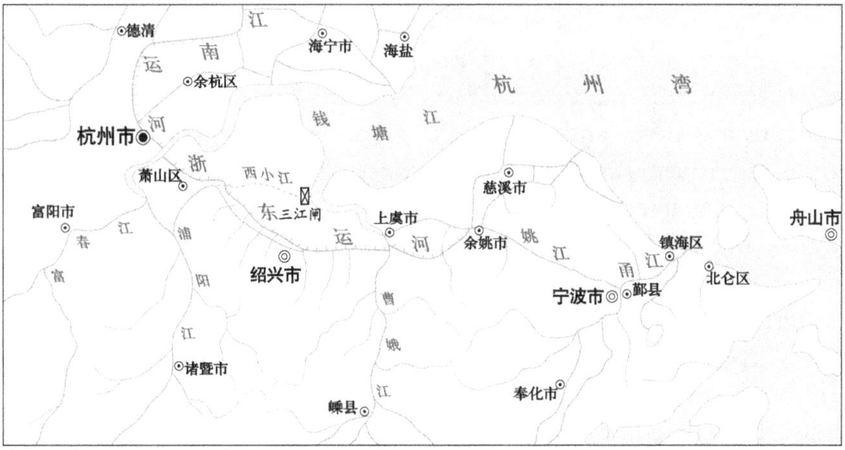

Figure 8.1: Map of the Zhedong Canal

of the Kuaiji and Siming Mountains in the south, approximately parallel to the coastline in the north, and runs through the piedmont plain from west to east. In terms of the regional topography, its elevation is higher in the south than in the north and slightly lower in the east than in the west. Generally, the elevation of the Xiaoshan–Shaoxing Plain ranges from 4.5 to 5.5 m. The Ningbo Plain, in the east, has a relatively low elevation. The elevation around Yuyao ranges from 2 to 2.5 m, with the lowest elevation between 1.5 and 2 m around Zhangting. There are low hills to the south of the plain with a height above sea level of less than 300 m. The elevation along the Zhedong Canal is between 4 and 6 m.

Affected significantly by the monsoon circulation, the Zhedong region has abundant rainfall. According to the hydrological statistics after the 1950s, the annual average rainfall varies between 1,100 and 1,300 mm in the Xiaoshan–Shaoxing–Ningbo Plain and is above 1,800 mm in the southern mountainous area. The rainfall has an uneven intra-annual distribution, concentrating mainly in the periods from May to June and from August to September. Generally speaking, the Zhedong region is relatively rich in water resources. The Zhedong Plain has abundant rainfall, a dense river network, and good conditions for navigation; the saying 'boats are vehicles and oars are horses' was coined here.

The main natural rivers in the region include the Puyang River (also known as the Qianqing River or the Xixiaojiang River in the downstream area), the Cao'e River, and the Yuyao River in the east. The Puyang River and the Cao'e River flow from south to north, while the Yuyao River flows from west to east. The three rivers all have the characteristics of mountainous rivers, with a great variation in their high and low water levels. Additionally, they are all tidal rivers affected significantly by the downstream tidal fluctuation. From the Eastern Han Dynasty to the Southern Song Dynasty, the Jianhu Lake was a key regulating reservoir in the area and an important water source for the Zhedong Canal. The functions of the lake, however, gradually disappeared due to a decrease in the lake's area after the artificial reclamation of part of it for agricultural farming during the Southern Song Dynasty.

The Zhedong Canal is 213.1 km long. It starts at Xixing Town, Xiaoshan City, across the Qiantang River from which is the Jiangnan Canal. The Zhedong Canal passes Shaoxing City and Ningbo City and flows into the sea at the mouth of the Yongjiang River. It is the extension of the southern end of the Beijing–Hangzhou Great Canal, thus having become one of the important waterways connecting the rivers around it and the sea in terms of navigation.

Due to how it was formed, the Zhedong Canal is divided into two parts; one is the artificial canal from Xixing to Tongming Dam (also named 'Jiangkou Dam') in Shangyu, with a length of 106.5 km, and the other is the natural river channel from Tongming Dam to the mouth of the Yongjiang River, with a length of 106.6 km. According to the features of the geographical environment and regional river system, the Zhedong Canal is divided into three relatively independent parts — the Xiaoshan–Shaoxing Canal, the Shangyu Canal, and the reaches of the Yuyao River and the Yongjiang River. Each of the three parts has different characteristics in physical geography and engineering technology.

The Xiaoshan–Shaoxing Canal is from Xixing to the Cao'e Dam, connecting the Qiantang River and the Cao'e River. The area around it is mostly flat and most of the natural rivers in it flow from the southwest to the northeast. The canal runs perpendicular to these rivers and, thus, a plain river network is formed with the canal as its backbone. In addition to navigation, the Xiaoshan–Shaoxing Canal has other important functions such as regional flood control, drainage, water delivery, and irrigation.

The Shangyu Canal connects the Cao'e River and the Yuyao River, and it was formed gradually by artificial improvement on the basis of natural rivers. The southern, northern, and central parts of the Shangyu Region are mountainous areas, and the canal flows through the plain between these mountains with a large change in topography and the complex flows of the natural river system. In terms of the characteristics of the topography, river system, and river channel, the Shangyu Canal has the features of a transitional reach between the artificial canal and the natural river. From the Ming and Qing dynasties onwards, the Shangyu Canal has become a double-route canal — the southern route is the traditional reach from Lianghu Dam to Tongming Dam, and the northern route is the new reach from Baiguan Town to Caoshu Bridge in Yuyao, with both the routes connecting the Cao'e River and the Yuyao River.

The part of the Yuyao River from Tongming Dam to Sanjiangkou and the part of the Yongjiang River downstream of Sanjiangkou are the natural reaches of the Zhedong Canal, which is also the main river channel to discharge floods in the Ningbo area. The Yuyao River has a distributary at Zhangting in an easterly direction, namely today's Cijiang River (also called 'Guanshan River' or the 'Zhongda River' in ancient times). The distributary and the downstream reach of the Yongjiang River converge in the east of Ningbo City. Artificial channels such as Cixijiang, Xiaoxibapu (also called 'Chazigang'), and Xitanghe were formed between natural rivers by river-channel improvement. There are water-control projects at the confluences of rivers.

This unique natural environment has resulted in unique technical characteristics in the Zhedong Canal projects. The Zhedong Canal integrates the rivers in the Zhedong region, forming a unified 'canal system' and playing a significant role in optimisation of the regional river network structure and water environment. Since the Southern and Northern dynasties, weirs have been used as the main control projects, and almost all of these projects are located at the confluences of the natural rivers and artificial canals.

8.2.2 *Historical Evolution*

8.2.2.1 *The Origin of the Canal*

The construction of the Zhedong Canal began in the Spring and Autumn Period, and the first project was the Shanyin Canal of the Yue Kingdom, which had a length of approximately 25 km from the eastern part of Shaoxing City to Cao'e. The initial purpose of the canal was not navigation. Later, with the enlargement of its scale, its functions were gradually widened as well, and the canal became an important navigational channel in the area.

In its early life, emphasis was placed on the improvement of local river reaches and the connection of natural rivers, with less large-scale excavation and project construction. From the Eastern Han Dynasty to the Northern and Southern dynasties, the tumultuous events caused a large number of people to move to the south, which promoted the development of Jiangsu and Zhejiang provinces and regional water transportation.

In the 5th year of the Yonghe Era in the Eastern Han Dynasty (AD 140), Ma Zhen, the satrap of Kuaiji County, led the construction of Jianhu Reservoir and a levee of 63.5 km between the Puyang River and the Cao'e River in Shaoxing. The reservoir, which had an area of 206 km^2, stored the water of 36 rivers originating from Kuaiji Mountain, supplying water for irrigation on the Shaoxing Plain.

Around AD 300, He Xun, the Manager of Kuaiji County in the Western Jin Dynasty, led the excavation of Xixing Canal as the irrigational trunk canal of the Jianhu reservoir. The Xixing Canal also functioned as a navigational canal, with a length of more than 100 km from Xiling (modern-day Xixing Town, Xiaoshan City) to Cao'e. The Xixing Canal ran parallel to the Jianhu Levee, with culverts on it to discharge floods into rivers. The Xixing Canal is the predecessor of the Zhedong Canal.

8.2.2.2 *The Preliminary Formation of the Zhedong Canal*

In the Northern and Southern dynasties, the application of many weirs and technology of towing ships over a weir to its downstream reach marked the preliminary formation of the basic characteristics of the Zhedong Canal. From AD 479 to 502, there were three weirs — Xiling, Beijin, and Nanjin — from west to east in the canal. The Xiling Weir (later called the 'Xixing Weir') was located at the confluence of the Xixing Canal and the Qiantang River. The Nanjin Weir and the Beijin Weirs (later known as the 'Qianqing Weirs') were respectively located on the eastern and western banks of the Puyang River at the intersection of the Xixing Canal and the Qiantang River. All three weirs went across the canal, controlling the waterheads at the confluences of the natural rivers and the canal to ensure an adequate water amount, depth, and velocity for the navigation of the canal. From AD 502 to 519 in the Southern Dynasty, Duci Weir (later named 'Dusi Weir') was built at the east of Shaoxing to control the water-discharge from Jianhu to the east by means of the canal.

During the Tang Dynasty, the Zhedong Canal was partially rebuilt in the west of Shaoxing City. The original canal went south around the city and a new canal was excavated in the north. Since Shaoxing City grew in size, the new canal flowed into the city at the west gate — named 'Ying'en' — out of it at the east gate — named 'Dusi Gate' — and then continued to flow eastward. The canal in the west of Shaoxing city was called Yundaotang (later known as 'Zhongtang'), and it flowed out of Shaoxing at its west gate and continued to flow across the Qianqing River until it arrived in Xiaoshan.

8.2.2.3 *The Formation of the Entire Zhedong Canal*

During the Song Dynasty, the entire Zhedong Canal (Figure 8.2), from Xiaoshan to Ningbo, was formed with preliminary water projects and management systems. The Zhedong Canal can be described as three reaches; the first reach was known as Xixing Canal and had a length of approximately 52.5 km, which started at Xixing town in Xiaoshan, flowed across the Qianqing River, and ended at Shaoxing

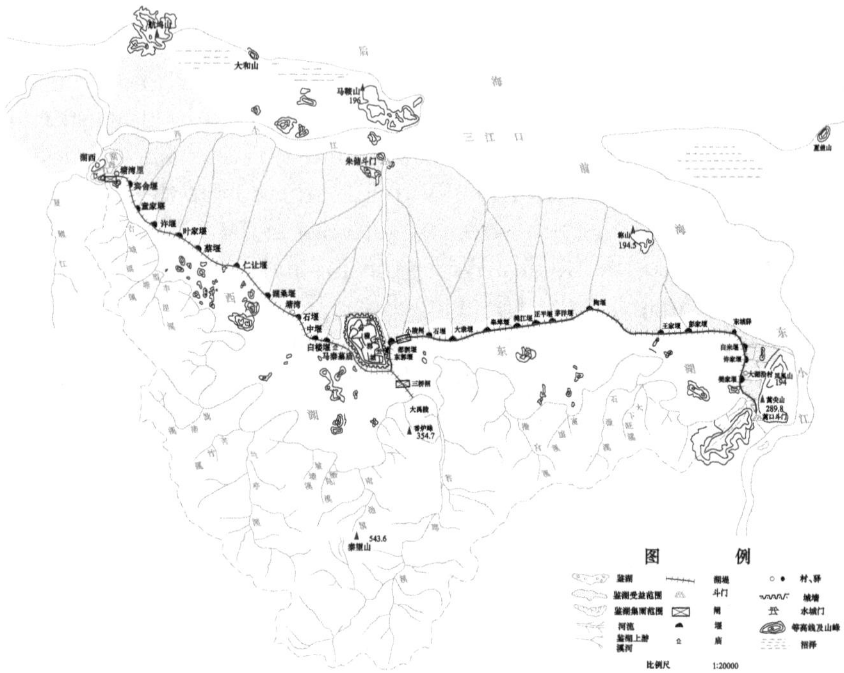

Figure 8.2: Schematic Diagram of the Zhedong Canal in the Song Dynasty

Notes: According to the illustration in *Research on the History of the Zhedong Canal*, written by Yao Hanyuan (with modification).

City. The second reach was approximately 50 km long, flowing out of the southeast of Shaoxing City, passing through Kuaiji County (east of modern-day Shaoxing City) and continuing to flow east until it arrived at the Cao'e River in Shangyu. The third reach had a length of approximately 15 km, flowing across the Cao'e River and finally entering the Yuyao River (also known as the 'Yao River'). The artificial canal was approximately 117.5 km long. The Yuyao River and the Yongjiang River, which were east of the artificial canal, were natural rivers. The complete navigation system of the Zhedong Canal was formed during this period, making it possible for ships to sail from the Qiangtang River to Ningbo City via Shaoxing City and then enter the sea at Ningbo City.

In the Southern Song Dynasty, Lin'an (modern-day Hangzhou) was the country's capital. The rich areas in the Zhedong region, such as Mingzhou (modern-day Ningbo), Shaoxing, and Taizhou, became the economic mainstays of the national government. As a result, the Zhedong Canal became an arterial shipping lane for the transportation of tribute rice. Meanwhile, the Zhedong Canal was a mainline for economic, political, and cultural exchanges between China and foreign countries, especially Korea and Japan. Most foreign envoys, traders, and monks landed at the Mingzhou Port and then went on to Central China through the Zhedong Canal. During the Southern Song Dynasty, the status of the Zhedong Canal was greatly enhanced. The national government managed the canal directly. The maintenance and dredging of every reach of the canal were carried out by special armies. Each weir was guarded and managed by 25 soldiers. This period was the age during which the Zhedong Canal flourished the most.

During the Song Dynasty, the water projects of the Zhedong Canal became more sophisticated than before. According to historical records, there had been seven weirs in the canal in this period. In a west-to-east order, the weirs were Northern Qianqing, Southern Qianqing, Dusi, Cao'e, Lianghu, Tongming, and Xidu.

In the 5th year of the reign of the Xining Emperor in the Northern Song Dynasty, a Japanese monk named Chengxun made a pilgrimage to China. He set out from Hangzhou and went along the Zhedong Canal to visit Tiantai Mountain. According to his records, after waiting for the tide to come in, his boat crossed the Qiantang River and entered the Xixing Canal directly. There was no weir in Xixing town, but there was a sluice foundation ruin at Dingqing Gate. There were two weirs at the confluence of the canal and the Qianqing River, where the river was more than 100 m wide, with a pontoon bridge across the river for pedestrians. The two weirs were named 'Northern Weir' and 'Southern Weir', respectively. The Northern Weir was close to the river, while the Southern Weir was 2.5 km away from the river. When he arrived in Shaoxing City, Chengxun's boat entered the city through Ying'en Gate and left it through Dusi Gate. There was a weir under Dusi Gate, and his boat was pulled over the weir by cattle. When he arrived at the Cao'e River, after waiting for the tide to come in, his boat was towed over the Cao'e Weir into the river. The boat sailed against the current in the Cao'e River until it arrived in Yan County. Then Chengxun walked to Tiantai Mountain.

At the beginning of the Southern Song Dynasty, the flow of the Qianqing River was fiercer, and because of river scour, the channel became deeper than it is today. According to historical records, the Qianqing River was more than 33 m wide and the water level in the canal was over 3.3 m higher than in the river. Therefore, there had to be two weirs on the river to control water and a pontoon bridge for pedestrians. Boats sailing on the canal were to be pulled over the two weirs to go through the Qianqing River, which was not only time-consuming and laborious, but also led to a grave safety hazard. In the 16th year of the Chunxi Era (AD 1189), when the remains of Emperor Gaozong were being transported from Hangzhou to Baoshan Tomb in Shaoxing, the fleet sailed on the Xixing Canal and crossed the Qianqing River on the way. At that time, there were two sluices in the canal that were temporarily built as control projects to replace the previous weirs. It was so dangerous for the fleet to go through the sluices that the officials of the escort for the fleet nearly committed suicide. In the 1st year of the Jiatai Era (AD 1201), two new weirs were built near the weirs in the Qiangqing River to increase the river's navigation capacity; at the same time, a new channel was excavated to extend the canal by 1 km.

The southern tributary of the upper Yao River was also known as 'Tongming River'. During the Northern Song Dynasty, a weir named 'Northern Tongming Weir' was built at the confluence of the canal and the Tongming River. The weir was 5 km east of the old Shangyu County (modern-day Fenghui Town). The Yao River had a steep bed slope due to the great impact of the tide, and the tide went out quickly. Large ships carrying heavy loads of salt were only able to cross the weir at high tide. When waiting for the tide to come in, the ships stayed on the canal and stopped other ships behind them from crossing the weirs. As a result, in the 1st year of the Jiatai Era (AD

1201), the Southern Tongming Weir was constructed 1.5 km east of the old Shangyu County. Then there were two confluences of the canal and the Yao River, and two canal mouths were formed at the northern and southern Tongming Weirs. The Northern Tongming Weir was used only by ships for the transportation of salt, while the Southern Tongming Weir was for official, civil, commercial, and tourist boats.

The Jianhu Reservoir had been gradually silted up since the Tang Dynasty, while reclamation was accelerated in the Song Dynasty. In the 15th year of the Jiatai Era in the Southern Song Dynasty (AD 1222), most of the reservoir's water surface had disappeared. The abandonment of the Jianhu Reservoir had changed the location of the water system on the Xiaoshan–Shaoxing Plain. Without the regulation of the Jianhu Reservoir for flood control, the Xixing Canal directly faced the risk of floods from southern mountains in the flood season, while it did not have enough stable water sources in the dry season. The Dusi Weir was originally built to keep water in the Jianhu Reservoir from flowing out of it to the east. After the abandonment of the Jianhu Reservoir, the Dusi Weir was abandoned as well.

8.2.2.4 *The Formation of a Double-Line Canal and Improvement of the Engineering System*

During the Ming and Qing dynasties, there was little change to the mainline of the Zhedong Canal compared with that in the Song Dynasty, but a new channel in the north of old Shangyu County was excavated to connect the Cao'e River and the Yuyao River. While the Tongming multi-purpose project was improved, another new channel — named 'Shibali' — was excavated. At Zhangting, in the middle Yuyao River, a distributary flowed east, turned south at Jiatianqiao in Cixi County, crossed the Yuyao River at Xidu Town, then passed through Gaoqiao along the Xitang River, and finally arrived at the west gate of Ningbo City (as shown in Figure 8.3).

The new canal in Shangyu was formed on the basis of improving the river in the north of Shangyu County. The new canal began at

Figure 8.3: Schematic Diagram of the Zhedong Canal in the Ming Dynasty

Notes: Based on the illustration in *Research on the History of the Zhedong Canal*, written by Yao Hanyuan (with modification).

Baiguan Town and flowed into the Cao'e River. Then it continued to flow east, crossing the Xiagai Lake and finally flowing into a natural river named the 'Jingjiang River'. The Jingjiang River was 7.5 km west of Yuyao County, flowing east into the Yuyao River at Caoshu Bridge, 10 km west of the Yuyao County.

There were three weirs in the new canal — namely Yiting, Henghe, and Doumen — and the new canal was parallel to the southern traditional canal. The reach connecting the Cao'e River and the Yuyao River had the worst natural conditions of the entire Zhedong Canal. At the beginning of the Qing Dynasty, there was a weir every 5 km in the canal from Lianghu to Tongming. Because of the difficulty of towing ships over the weirs, big ships could only carry dozens of dan (1 dan is equal to 0.1 m^3) of goods, but hundreds in the Song Dynasty. Therefore, more ships sailed on the northern line than on the southern one at the beginning of the Qing Dynasty.

After running through Yuyao County, the Yuyao River bifurcated into two branches at Zhangting; one was the south-eastern branch, the main navigation canal, which flowed into the Yong River and then into the sea in the east of Mingzhou (modern-day Ningbo City), while the other was named "Guanshanjiang River' or 'Zhongdahe River' (now known as 'Cijiang River'), which flowed east through the southern part of Cixi County (modern-day Cixi Town) and bifurcated again at Jiantian Bridge, with one branch flowing south into the main stream of the Yuyao River and the other branch continuing to flow east until it reached Zhenhai County. The original purpose of the Guanshanjiang River was to divert water from the Yuyao River to irrigate Cixi and Zhenhai counties, but it also aided navigation (see Figure 8.3).

During the Ming Dynasty, the Puyang River changed its course at Qiyan and then flowed west into the Qiantang River. In the 2nd year of the Chenghua Era (AD 1466), the Qianqing weir was abandoned. The fact that the Puyang River changed its course largely altered the location of water resources in the Xiaoshan–Shaoxing Plain. On the one hand, flooding and waterlogging disasters that had previously been caused by the Puyang River were alleviated on the plain, but on the other hand, the shortage of water sources and the influence of tide became serious. The Xixiaojiang River was no longer a channel to discharge floods from the Puyang River, but it continued to be a drainage channel for the Xiaoshan–Shaoxing plain and a water source for irrigation. The reduction of stream flows resulted in serious sedimentation in the Xixiaojiang River, especially in its lower reaches. The river was wide at its estuary and no floodwater flowed into the sea, all of which resulted in seawater intrusion. As a large amount of brackish water flowed back into the river with sediment, the sedimentation in its estuary became ever more serious. Because of the sedimentation and the tidal backwater in the estuary, pluvial waters could not be drained out of the Xiaoshan–Shaoxing Plain, while the seawater intrusion led to the shortage of fresh water for irrigation. A lack of water-control projects meant the water level in the canal changed frequently with the water level in the Xixiaojiang River, which was quite unfavourable for navigation.

In the 14th year of the reign of Jiajing (AD 1535), Shao En, a new Magistrate in the Shaoxing Prefecture, led the construction of a sluice at Sanjiangkou, where the Xixiangjiang flowed into the Hangzhou Bay. The sluice was named 'Sanjiang' and had 28 openings. Seawalls were also built on both sides of the sluice, and some stage-gauging stations — known as 'stage-gauging stones' — were set up beside the sluice and in Shaoxing City to mark water levels by the five elements — 'metal, wood, water, fire, and earth'. The opening and closing of the sluice was strictly managed by means of the established standard water levels. After it was completed, the Sanjiangzha became the core hub project in terms of controlling the water level of the river network on the Xiaoshan–Shaoxing Plain. Ordinarily, the sluice was closed to stop the tide from flowing into the river, but during the rainy season it was left open to discharge floodwater from the river. The sluice also had the functions of irrigation and navigation. As a result, a unified water system known as the 'Sanjiang Water System' was formed, including the canal in the Xiaoshan–Shaoxing Plain. Under the new conditions that the Puyang River changed its course and the Xixiaojiang dried up, the completion of the Sanjiang Sluice provided the water and correct water level for the canal, thereby also providing stable conditions for its operation.

8.3 Key Projects

The topography, water system, and hydrological characteristics of the Zhedong Canal have determined the features of its water-control projects. The overall longitudinal gradient of the canal is gentle, and the topographical conditions of the area are good. The Zhedong Canal intersects with many natural rivers, most of which are mountainous rivers and tidal rivers. As a result, the water levels and discharge along the canal change greatly and frequently. The water-control projects along the Zhedong Canal are located at the confluences of artificial channels and natural rivers. In the Ming and Qing dynasties, the Sanjiang Sluice was a water-control hub project of the river network in the Xiaoshan–Shaoxing Plain, with the Zhedong Canal as its backbone. The use of weirs as water-control projects over a long period is a significant feature of the Zhedong Canal.

8.3.1 *Navigation Projects*

8.3.1.1 *The Main and Branch Lines of the Canal*

The main route of the Zhedong Canal had a total length of 213.1 km. The reach from the east bank of Qiantang River at Xixing Town, Xiaoshan County, Hangzhou, to the Big Jiangkou Weir in Shangyu County is an artificial channel, and the one from the Big Jiangkou Weir to the mouth of the Yongjiang River in Ningbo consists mainly of natural channels. The canal in Shangyu County (see Figure 8.4) and Cixi County is composed of double lines, each of which was the main navigation channel in a certain historical period. Their excavation was intended to enhance the navigational ability of the Zhedong Canal. In addition to the main route, there are some artificial branch lines that flow from the main route to the north or to the south. Haoba Canal, for example, was originally a channel connecting the

Figure 8.4: Current Location of the Northern Shangyu Canal (2010)

Figure 8.5: Current Location of the Northern Moat of Yuyao County (2010)

eastern side of the Jianhu Lake and the Cao'e River. It later became a canal that flowed south to Sheng County, and thus it was once called the 'Canal of Sheng County'. The Chazigang and Xitanghe rivers, both of which had a trend from north to south, connected the east-to-west navigational channels of the Guanshan River (see Figure 8.5), the Yuyao River, and the moat of Ningbo City. As a result, a crisscross of regional river navigation networks was formed.

The water system of the Zhedong Canal has been gradually formed through constant expansion and improvement in the process of historical development. In the evolutionary process of the Zhedong Canal, the planning of each period made full use of the regional geographical conditions and the characteristics of the water system. A continuous and optimised canal, namely the Zhedong Canal, was finally formed, composed of artificial and natural channels and connecting inland canals and the sea for navigation. Due to its differing features of topography and differing water systems, each reach of the canal also presents different engineering characteristics. Because most of the natural rivers have a trend from north to south in the Xiaoshan and Shaoxing areas, it is necessary to excavate canals from east to west for navigation. In the east of Shangyu County, the canal made full use of the natural rivers that trended from east to west. With the reach in Shangyu County as a transitional channel, a continuous canal connected to the sea was formed. The elevation of the artificial reaches of the Zhedong Canal was controlled so that it ranged from 4 to 6 m, with an average longitudinal slope of less than 0.01 per cent. This allowed the canal to have a steady flow, an adequate amount of water, and an unblocked channel. The average longitudinal slope of the natural reaches of the Zhedong Canal was 4.71 per cent. With the reach in Shangyu County as a transitional channel and especially with the Tongming Weir, a hub water project, as a control project, the artificial and natural channels were well-connected, which has ensured the long-term, stable operation of the canal for navigation. The north of the canal is far away from the coast, which makes it possible for the canal to avoid the intrusion of the tide and avoid the impact of the gradual change in the Qiantang Bay, ensuring the safety of the canal. The Zhedong Canal is one of the reaches that has been continuously operated for the longest time, having the least problems and the most stable operation along the entire Grand Canal. On one hand, this is attributed to the favourable natural and geographical conditions of the Zhedong region, but on the other hand, it can also be attributed to the scientific and reasonable planning of the canal.

8.3.1.2 *Canal Levees as Towpaths*

There are dense river networks in the Zhedong Plain, especially in the Shaoxing area. The canal and rivers converge on the plain. A large number of bridges, culverts, and small sluices — '*doumen*' in

Figure 8.6: Ancient Towpaths of the Shaoxing Canal (2012)

Chinese — were built on the levees of the canal. The levees also served as towpaths (see Figures 8.6 and 8.7).

Since its completion in the Yonghe Era in the Eastern Han Dynasty (AD 136–141), the Jianhu Lake had always been an important regulating reservoir for the Xiaoshan–Shaoxing Plain until it was abandoned due to artificial reclamation during the Southern Song Dynasty. Around AD 300, He Xun, a Manager of Kuaiji Prefecture at the time, led the excavation of the Xixing Canal, which was adjacent to the levee of the Jianhu Lake. On the levee of the Jianghu Lake, a large number of culverts and bridges were built, serving as passages between the canal and the lake (see Figure 8.8). After the Jianhu Lake was abandoned during the Jiajing Era in the Ming Dynasty, Tang Shaoen led the reconstruction of the southern levee of the Shaoxing Canal in order to improve the water exchange between the vestige of the Jianhu Lake and the canal. As a result, most of the levees of the Zhedong Canal were water projects with multiple functions, such as keeping water

Figure 8.7: Ancient Towpaths of the Xixing Canal in Xiaoshan County (2010)

Figure 8.8: Ruins of the Bridges between Jianhu Lake and the Canal (2010)

from flowing out of the canal, connecting the canal and the lake, and serving as a towpath (see Figures 8.6 and 8.7).

8.3.1.3 *The Relationship Between the Canal and Cities*

The Zhedong Canal runs through cities such as Xiaoshan, Shaoxing (see Figure 8.9) (including the historical Shanyin and Kuaiji Counties), Shangyu, Yuyao, Cixi, and Ningbo (see Figure 8.10). The moat systems in these cities had a close relationship with the canal — either the canal ran through the cities or its branches entered them. The reaches of the canal in the cities were an important part of their moats. Each city had some water gates of its own to control boats that came into or out of the city. Meanwhile, moats connected to the canal had the functions of draining urban pluvial water and supplying water for the urban environment.

Shaoxing City was the core of the Zhedong region. It was also typical of how the canal was closely related to the cities. There was a dense river network in the Shaoxing area, and boats were the main transportation tools. The canal was the backbone of the east–west rivers in the plain, and the channels in and around the city served as the centre of the canal.

Shaoxing City was originally built in the Spring and Autumn Period, and it was reconstructed and extended later on. It had a perimeter of more than 12 km during the Song Dynasty. After being extended in the Yuan Dynasty, Shaoxing City had five gates reserved for land transportation, two gates for both land transportation and navigation, and four gates reserved for navigation. This layout continued until the Ming and Qing dynasties, but the name of each gate changed. The main route of the canal entered the city at Ying'en Gate and left it at Dusi Gate. The river around the city was connected to the canal and there was a crisscross of river networks inside the city, which formed a water system with such functions as navigation, water supply and drainage, and improvement of the urban landscape. Each channel in the city was connected either to the canals (see Figure 8.9).

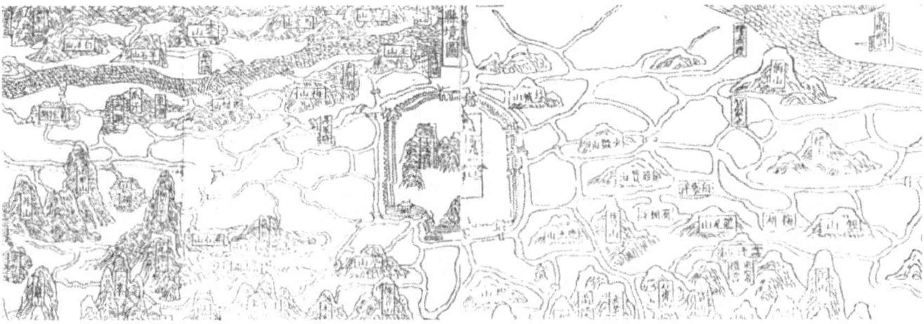

(a) Relationship between Shaoxing City and Canal (According to "Wanli Period" Local Chronicles of Shaoxing City)

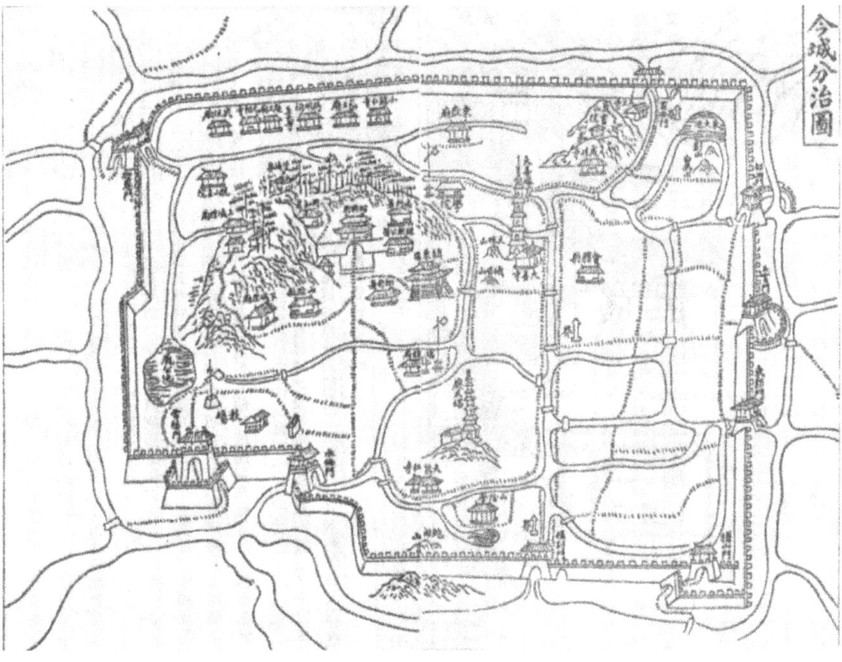

(b) Relationship between river system in the city and the canal(According to"Jiaqing Period"Local Chronicles of Shanyin County)

Figure 8.9: Relationship Between the Moat System of Shaoxing City and the Canal

Figure 8.10: The Confluence of Three Rivers in Ningbo City (2010)

8.3.2 *Projects at the Confluences of the Canal and Natural Rivers*

8.3.2.1 *Xixing Weir/Sluice — A Project at the Confluence of the Canal and the Qiantang River*

The mouth of the Xixing Canal is located at the western end of the Zhedong Canal, serving as a gateway between the eastern and western parts of Zhejiang. Xixing Town was originally called 'Guling', and later it was known as 'Xiling' and 'Xixing'. It has been a gateway for both land and water transportation in the Zhedong region, taking an important strategic position in terms of military roles and economic exchange.

Hangzhou City is across the Qiantang River from Xixing Town. Hangzhou was the country's capital in the Southern Song Dynasty. Marked by the construction of Xiling Weir in the Southern and

Northern Dynasty, the Xixing canal mouth project (a water hub project) was formed.

The evolution of the Xixing canal mouth project featured the following stages: formation and perfection in the 4th to 5th century, a booming period in the 5th to 13th century, gradual decline in the 14th to 16th century, and gradual abandonment after the 16th century. The type of the control project at the canal mouth also went through several changes (see Table 8.1). The operational method of shipping at the canal mouth and its main functions experienced the following stages: before the 14th century, boats sailed through the

Table 8.1: The Evolution of Control Projects on the Xixing Canal

Time	Name	Profile of the project
The Southern and Northern Dynasties (AD 420–589)	Xiling Weir	The only way of passing through the states of Wu and Yu; the boats were to be pulled over the weir in the canal mouth by cows.
The last years of the Southern Chen Dynasty (AD 583–589)	Fenggong Weir	Also known as 'Fenggong Weir'.
Tang Dynasty (AD 618–907)	Xiling Weir	Called 'Xiling' again.
The Reign of Qianliu in the Wuyue State (AD 902–932)	Xixing Weir	Qianliu stationed troops here and changed the name to 'Xixing' because he didn't like the word 'Ling', meaning 'tomb'.
The last years of the Five Dynasties to the early years of the Northern Song Dynasty	Xixing Sluice	The weir was a rebuilt sluice to aid shipping.
The middle and later years of the Northern Song Dynasty	Xixing Sluice	The sluice was abandoned because shipping no longer required it.

(*Continued*)

Table 8.1: (*Continued*)

Time	Name	Profile of the project
The early years of the Southern Song Dynasty	Two sluices at Xixing connected the Qiantang River	Two sluices were built in the canal mouth to intercept sand flowing into canal with the tide from Qiantang River. The idea was to remove sedimentation from the canal bed.
The early years of the Ming Dynasty	Big Weir	Because of the poor results of the previous sluice, a weir was rebuilt instead to thoroughly solve the canal sedimentation.
The 15th year of Wanli Era in the Ming Dynasty (AD 1587)	Yongxing Sluice (or 'Dragon Mouth Sluice')	Liu Hui, who was the County Magistrate of Xiaoshan, reconstructed the sluice instead of the Big Weir to aid drainage.
The 58th year of the Qianlong Era in the Qing Dynasty (AD 1674)	Yongxing Sluice (or Dragon Mouth Sluice)	The sluice project was rebuilt by the new County Magistrate, Jiang Chongyao, this year.
Before the 24th year of The Republic of China (Before 1935)	Yongxing Sluice	To clean the sluice floor and its surroundings, and to change the flashboard so it diverted water from the Qiantang River to combat the drought.

sluice directly or were towed over the weir; from the 14th to the 16th century, boats could only travel along the canal and goods or people in the boats had to be transferred to other boats to cross the Qiantang River; and in the late 16th century, the main function of the canal mouth was gradually changed from navigation to regional drainage.

The Xixing Hub Project consisted of control projects such as weirs and/or sluices, the canal inside the project and the sandy channel outside the project (see Figure 8.11). The control projects were the core of the hub and also an integral part of the seawall of the Qiantang Bay, and their locations were relatively fixed. However, the length and shape of the sandy channel were significantly influenced by

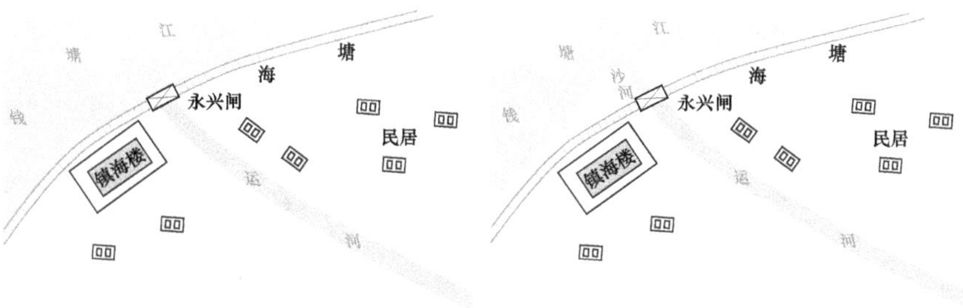

(a) Without sandy channel (b) With sandy channel

Figure 8.11: Schematic Diagram of the Relationship between the Xixing Canal, the Seawall, and the Qiantang River

changes in the Qiantang River regime. After the Qianlong Era in the Qing Dynasty, the sandy channel became unstable due to the frequent change in the Qiantang River regime, which eventually resulted in the abandonment of the canal mouth project on the Xixing Canal.

8.3.2.2 *Qianqing Weirs, Projects at the Confluence of the Canal and the Puyang River*

The two weirs at Qianqing (known as the 'Qianqing Weirs' for short) were control projects located at the confluence of the canal and the Puyang River (before it changed its course). The Puyang River was the biggest river with a south–north trend, running through the Xiaoshan–Shaoxing plain. It had a rapid flow and there was a big difference between the water level in flood season and that in the dry season. Therefore, as water-control projects, weirs were built at Qianqing Town, which was the confluence of the canal and the Puyang River. There were two weirs at Qianqing that were on the canal, but they were located on both sides of the Puyang River.

According to the records of Chengxun in the 5th year of Xining Era in the Northern Song Dynasty (AD 1071), the northern weir at Qianqing was adjacent to one of the two banks of the Puyang River, while the southern weir at Qianqing was 2.5 km away from the other bank of the Puyang River. The Puyang River had a width of 100 m, across which there was a pontoon bridge supported by 10 small boats. Afterwards, two more weirs were built near the old ones to increase

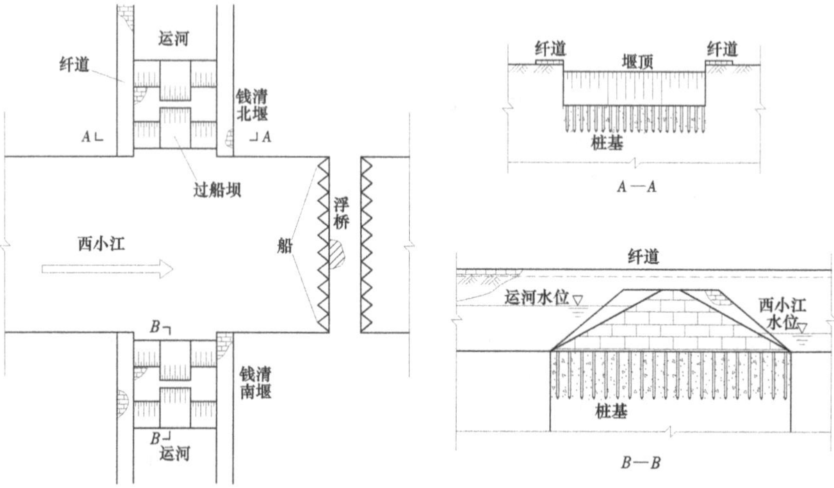

Figure 8.12: Suggested Layout and Structure of the Qianqing Weirs before the Chenghua Era in the Ming Dynasty

the navigation capacity. Two small sluices were built once to regulate water in the canal, but they were not there for a long time.

After the Xuande Era in the Ming Dynasty, the lower Puyan River gradually changed its course and flowed into the Qiantang River via Qiyan Weir. As the Xixiaojiang River was gradually silted up, the Qianqing Weirs were demolished in the 2nd year of the Chenghua Era (AD 1466) due to their impediment to sailing boats. The Qianqing Weirs were hub projects at the confluence of the canal and the Puyang River in that period, with main functions such as controlling the water level of the canal, keeping water in the canal, stopping the tide from flowing into the canal, and promoting navigation. The layout and structure of the Qianqing Weirs are inferred as shown in Figure 8.12.

8.3.2.3 *Cao'e Weir and Lianghu Weir, Water-Control Projects at the Confluence of the Canal and the Cao'e River*

Historically, the Cao'e River and the Puyang River were known as 'East Small River' and 'West Small River', respectively. Both of them flow around Shaoxing City and then into the Hangzhou Bay. The confluence of the three rivers is known as 'Sanjiangkou'. The Cao'e River is typical of mountainous and tidal rivers. Cao'e Weir and Lianghu Weir were projects that enabled the canal to cross the Cao'e River, and their layouts are similar to that of the Qianqing Weirs. Located on the west and the east bank of the Cao'e River respectively, Cao'e Weir and Lianghu Weir were first built in the Southern and Northern dynasties to control the water level in the canal. During the Song Dynasty, a sluice was built beside Cao'e Weir and there was also a doumen near Cao'e Weir, both of which were used to regulate the amount of water in the canal. During the Wanli Era in the Ming Dynasty, both the sluice and the doumen were abandoned. Cao'e Weir, on the west bank of the Cao'e River, had a relatively fixed location, while the location of the weir on the east bank changed frequently due to changes in the location of the canal mouth on the east bank. In the Jiajing Era, the west bank of the Cao'e River near the canal was eroded, while the east bank was silted up and the width of the sand increased by 3.5 km. Zheng Yun, a magistrate in Shangyu Prefecture at the time, led the excavation of a channel on the sand and moved the weir on the east bank to the side of the river.

8.3.2.4 *Tongming Weirs, Projects at the Confluence of the Canal and the Upper Yuyao River*

With the Tongming Weirs as a boundary, the Zhedong Canal was divided into two reaches. One was the artificial reach west of the Tongming Weirs, and the other was the natural reach east of them, with the Yuyao River and the Yongjiang River as its main parts. Most natural rivers on the western plain of the Zhedong region flow from south to north. As a result, the channels with an east–west trend were all artificial canals. The eastern reach of the Zhedong Canal made full use of the natural rivers — the Yuyao River and the Yongjiang River — which trended from east to west. The natural rivers and the artificial canals in the Xiaoshan–Shaoxing plain constituted a continuous waterway that ran through the Zhedong Plain and finally into the sea. The average longitudinal slope of the artificial reach of the Zhedong Canal was less than 0.01 per cent and its water level was basically stable. As the main line of the Zhedong Canal, the main streams of the Yuyao River and the Yongjiang River were used to discharge floodwater from the Ningbo area. The two rivers had an average longitudinal slope of 4.71 per cent and a big difference between their water levels in the flood season and dry season. They were significantly affected by the tide at the Yongjiang River mouth, which could go upstream as far as the Tongming Weirs at Shangyu. With completely different hydrological features, the two reaches (the artificial reach and natural reach) of the Zhedong Canal were well-connected by the Tongming Weirs, which ensured a long-term smooth operation of the entire Zhedong Canal.

In the Northern Song Dynasty, the Tongming hub project was only a weir east of Shangyu County. To the west of the Tongming Weir was the artificial canal leading to Lianghu Weir, while to the east of the Tongming Weir was the Tongming River. In the Southern Song Dynasty, in order to enhance the navigational capacity of the canal, one more weir was built at Tongming and the Yue River was excavated. From then on, the two weirs at Tongming were called 'North Tongming Weir' and 'South Tongming Weir' respectively. During the Ming Dynasty, a large-scale reconstruction of the Tongming Project was carried out. The North Tongming Weir was moved to a place near Zhengjian Mountain. Two canals named 'Houxin' and 'Shibali' were excavated, and Dajiangkou Weir was built. The Tongming Hub Project was finally formed, consisting of three weirs and two shipping lanes. As a result, the navigational conditions and reliability of the canal were greatly enhanced. During the Qing Dynasty, due to lack of maintenance of the southern line in the Shangyu reach of the canal, its navigational conditions deteriorated and the northern line from Baiguan Town to Caoshu Bridge, west of Yuyao County, gradually came into use. The northern line once replaced the southern line as the main line of the Zhedong Canal.

As a project connecting the artificial channels and the natural rivers, the Tongming Hub Project consisted of weirs and the Yue River after it was rebuilt several times. Initially, there was only the North Tongming Weir, and the shipping lane at that time was the line marked by the numbers '1-1-1-1-2' as shown in Figure 8.13. In the 1st year of the

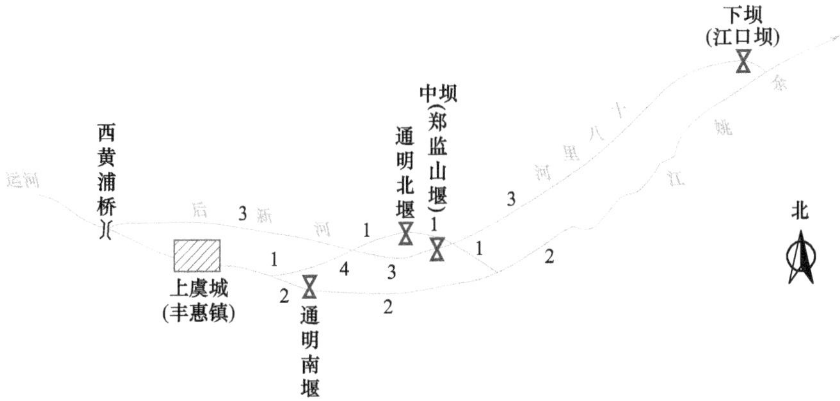

Figure 8.13: Evolution of the Canal Around Tongming Weir in the Ming Dynasty

Jiatai Era in the Southern Song Dynasty, the South Tongming Weir was built 1.5 km east of Shangyu County. After the completion of the Southern Tongming Weir, both the north shipping lane (marked by the numbers '1-1-1-1-2') and the south shipping lane ('2-2-2') were used.

In the Hongwu Era in the Ming Dynasty, the North Tongming Weir was moved to the foot of the Zhengjian Mountain, and therefore the new weir was named 'Zhengjian Mountain Weir', which was also known as 'New Tongming Weir' or 'Middle Weir'. After the New Tongming Weir was built, the shipping lane became the line marked by the numbers '1-4-3-1-2' as shown in Figure 8.14.

Qilitan was 2 km east of the South Tongming Weir, where the river was shallow due to sedimentation and sailing boats often had to wait for the tide to come in. In the 9th year of the Yongle Era (AD 1524), the shipping lane became obstructed due to serious sedimentation at Qilitan, and therefore a new canal, 'Houxin', was excavated from the West Huangpu Bridge (approximately 1.5 km west of the county) to Zhengjian Mountain Weir. The Houxin Canal was then extended 9 km eastwards Eld the Yuyao River; at their confluence, a weir named 'Jiangkou' (also known as the 'Xia Weir') was built. After the Yongle Era, official and civilian ships sailed along the line marked by the numbers '3-3-3' and merchant ships sailed along the shipping lane through the southern weir.

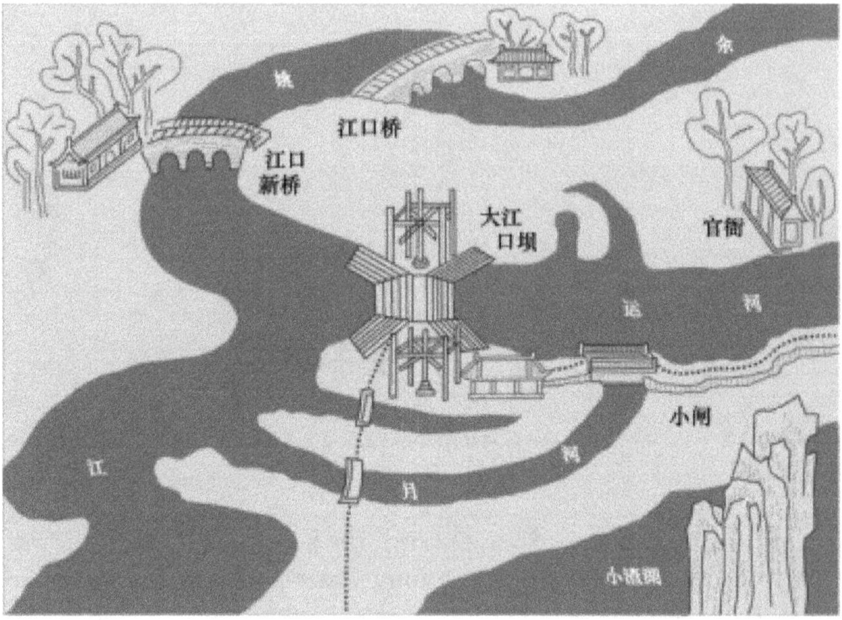

Figure 8.14: Confluence of the Zhedong Canal and the Yuyao River

Note: Based on illustration in *(Wanli Period) Local Chronicles of Shaoxing City* (with modification).

The Jiangkou Weir was also called the 'Lower Tongming Weir', 'Big Jiangkou Weir' or the 'New Weir'. It was situated within Yuyao County, 20 km northwest of the capital city of Yuyao County and 9 km southwest of the Middle Weir (see Figure 8.14). Jiangkou Weir was made of masonry, and the water level in the canal was 5 m higher than in the Yuyao River. Jiangkou Weir was the only route through which boats could pass from Shaoxing to Ningbo. As Jingkou Weir was quite high, it was difficult for boats to pass it, so the boats tended to wait for the tide to come in at night.

8.3.2.5 *The Large West Weir and Small West Weir, Projects at the Confluence of the Canal and the Lower Yuyao River*

Initially constructed in the Song Dynasty, Xidu Hub Project was a control project at the confluence of the Yuyao River and the

Chazipu–Xitanghe reach of the branch canal that connected Cixi County (modern-day Cixi Town) and Ningbo City. Outside the West Gate of Ningbo City was the Xitanghe Canal, which was connected to the main stream of the Yuyao River at Xidu Town.

The Yuyao River bifurcated into two rivers at Zhangting. One was its main stream, which flowed southeast to Ningbo, passed the north of Ningbo City, and then continued to flow east into the Yongjiang River and finally joined the sea. The other flowed east from Zhangting to the south of the capital city of Cixi County, joined Chazipu Canal at Jiatian Bridge, and then flowed south into the sea at Xidu.

At the confluence of the artificial branch canal and the main stream of the Yuyao River, there were control weirs — namely the Xidu Hub Project, consisting of the Large West Weir and the Small West Weir. The Xitanghe Canal and a weir at Xidu appeared no later than the Song Dynasty. In the Ming and Qing dynasties, the weir at Xidu was known as the 'Large West Weir' (see Figure 8.15). The Chazipu Canal on the northern bank of the Yuyao River had been formed during that period, and at the confluence of the Chazipu

Figure 8.15: Ruins of the Large West Weir (2010)

Canal and the Yuyao River, there was a weir named the 'Small West Weir'. Across the Yuyao River from there was the Large West Weir. Its main functions were resisting the tide, regulating water, and shipping.

8.3.3 *Sanjiang Sluice*

Located at Gusanjiangkou, on the southern bank of Qiantang Bay in the Northern Shaoxing Plain, the Sanjiang Sluice (see Figure 8.16) was built in the 16th year of the Jiajing Era in the Ming Dynasty (AD 1537). It is the largest of the existing multi-opening masonry sluices. It was also a key control project in the Zhedong Canal during the Ming and Qing dynasties, holding the functions of controlling the water levels in the canal, resisting brackish water from the sea, and storing fresh water in the canal.

After Qianqing Weir was abandoned, the Xiaoshao Canal did not have any key projects to control the quantity and level of water in the canal. The construction of the Sanjiang Sluice made it possible to

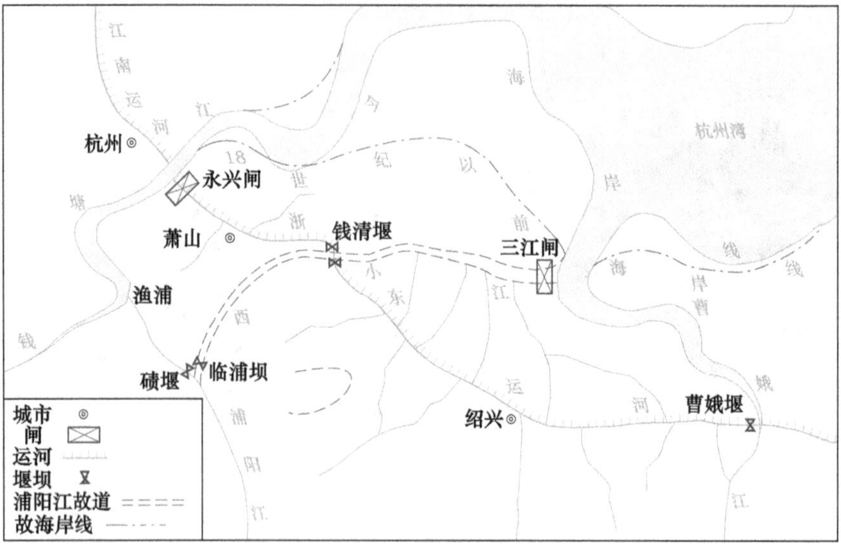

Figure 8.16: Schematic Diagram of the River System and Control Projects on the Xiaoshan–Shaoxing Plain in the Ming and Qing Dynasties

Figure 8.17: Current Status of Sanjiang Sluice (2010)

prevent seawater intrusion, to drain pluvial water smoothly, and to ensure stable water sources for navigation and irrigation. After the Sanjiang Sluice was built, the amount and level of water in the canal and the Qianqing River could be controlled. When water levels were low in the canal, the sluice was closed to store fresh water. When heavy rains and/or large floods resulted in high water levels in the canal, the sluice was opened to discharge floodwater. When the tide came in, the sluice was closed to stop it from flowing into the canal. As a result, the Sanjiang Sluice ensured good conditions for navigation along the Zhedong Canal. Four hundred years have passed since the Sanjiang Sluice was built, but it still functions normally (see Figure 8.17), was built, but it still functions normally, which is a tribute to the scientific consideration of its location, construction, and operation.

After the Puyang River changed its course and Qianqing Weir was abandoned, the Zhedong Canal was controlled by the tide at Sanjiangkou, and therefore its navigational conditions deteriorated. Originally, the Xixiaojiang was controlled by the law of stream flows, but it was completely controlled by the law of the tide at Sanjiangkou after the Puyang River changed its course. Although it was no longer a channel that discharged floodwater from the Puyang River, the Xixiaojiang River was still a main channel for drainage in the Xiaoshan–Shaoxing Plain and also had the function of irrigation. However, because of the riverbed aggradation and impact of the tide, as well as a lack of control projects, it was difficult to discharge floodwater in the region through the Xixiaojiang River. Meanwhile, due to inadequate stream flows in the Xixiaojiang River, brackish water flowed into the river, even flowing as far as Shaoxing City, which resulted in a shortage of fresh water and the salinisation of farmlands, as well as the serious deterioration of the water environment in the Xiaoshan–Shaoxing Plain.

The XiaoShao Canal was the east–west artery connecting rivers in the Xiaoshan–Shaoxin Plain. After Qingqian Weir was abandoned, the canal reach from Xixing to Cao'e Weir, which had a length of more than 100 km, was connected directly to the Xixiaojiang River. As a consequence, the canal changed from being a relatively independent channel with relatively stable amount and level of water to being controlled by the natural law of the tide at Sanjiangkou. At the same time, the Xixiaojiang River lost the balance between storing and discharging water, and seawater intrusion resulted in the deterioration of the regional water environment. These issues were extended to the whole Xiaoshan–Shaoxing Plain through the canal system. Due to the unbalance between storing and discharging water and the impact of the tide, the water quantity, water level, and flow in the Xixiaojiang River became unstable. The navigational conditions of the Xiaoshao Canal deteriorated rapidly because the Puyang River changed its course, and Qianqing Weir was abandoned. After it was built, Sanjiang Sluice played a crucial role in resisting the tide, storing fresh water, and discharging floodwater while at the same time achieving the goal of controlling the navigational conditions of the Xiaoshao Canal. It replaced Qianqing Weir and became a key control project on the Zhedong Canal from the 16th century to the 19th century. It has great scientific value in terms of its planning, design, construction, and management.

Sanjiang Sluice is located at Sanjiangkou, a vital point that connects the sea and the rivers in the Xiaoshan–Shaoxing Plain. The reason why this location was chosen is that it could effectively and completely control the operation of storing water and discharging water in and from the rivers and could prevent seawater intrusion.

The foundation of Sanjiang Sluice is fresh bedrock. Living stones were embedded in the bedrock by piercing holes in it, and they were then fixed by pouring liquid iron on top. Large slabs were paved on the living stones. The sluice was made of boulders, each of which weighed more than half a ton. Close attention was paid to the design of its structure and the construction method (see Figure 8.18). A flume was built on each side of the sluice piers. Stone sills were paved on the bottom of the sluice. Slabs were paved between adjacent piers as a bridge for traffic. The size of the sluice (as measured in 1935) is as follows: the length of the sluice is 103.15 m, with 28 gates, and the net width of the sluice is 62.74 m. The bottom elevations of the gates depend on natural rock foundation; the maximum depth of a gate is 5.14 m, while the minimum is 3.4 m. The net width of each gate is

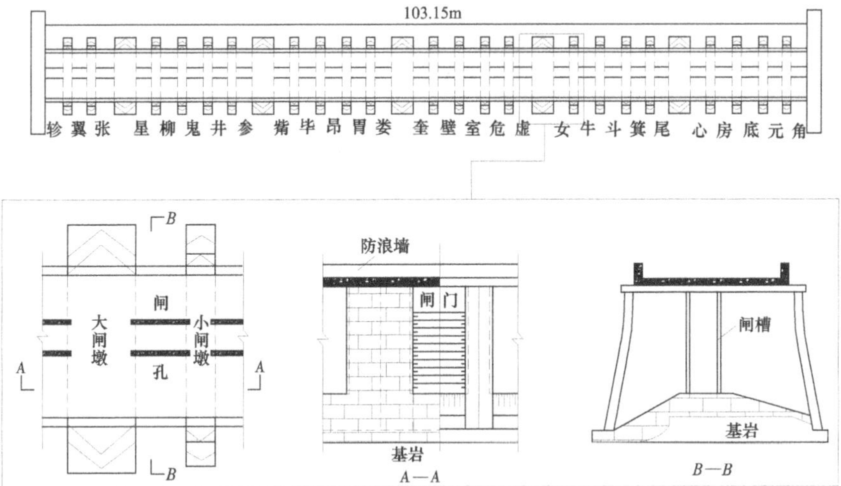

Figure 8.18: Schematic Diagram of the Structure of Sanjiang Sluice

Notes: Some data is from the *Report on Construction of Shaoxing Sanjiang Sluice* by Dong Kaizhang and the monthly journal *Water Resources*, Vol. 5, 1933: 1.

different, with a maximum of 2.41 m and a minimum of 2.16 m. The elevation of the sluice crest is 8.5 m above sea level at Wusong, and the maximum discharge of the entire sluice is 384 m^3/s.

By using the stage-gauging stelas and considering the requirements of protection against the tide, flood discharge, irrigation, and navigation, Sanjiang Sluice took quantitative control of the water level and quantity of water in the Xiaoshan–Shaoxing river network with the canal as its artery.

Shortly after the construction of Sanjiang Sluice, the Shanyin–Kuaiji stage-gauging stela (set up at Yousheng Abbey by Dai Hu in the Chenghua Era) was directly moved to the area near Sanjiang Sluice. The stage-gauging stela consisted of upper, middle, and lower scales. It was specified that when the rice seedlings in low-lying paddy fields were growing strong but wheat and vegetables had not yet been harvested, the water level should be controlled so that it could be more than a sixth higher than the lower scale, or else shipping would be impeded.

In the Wanli Era, two 'Sanjiang Stage-Gauging Stelas' (see Figure 8.19) were specially manufactured with the five words 'metal, wood, water, fire, earth' in Chinese as scales from top to bottom on the stone. The two stelas were set up in the river near the sluice and in the channel inside the city. A scientific and strict management system was formulated and the operators of the sluice were asked to use it according to the rules inscribed on the stelas as follows:

> When water reached the bottom of the word 'metal', all the gates have to be opened; to the bottom of 'wood', sixteen gates; and to the bottom of 'water', eight gates. If water level is above the top of 'fire' in summer or above the top of 'earth' in autumn, all the gates have to be closed tight. The operators of the sluice had to follow the rules strictly and timely.

As a result, unified regulation was carried out across the entire river network in the plain, and the goal of regulating water quantity was accomplished through the quantitative control of the water levels.

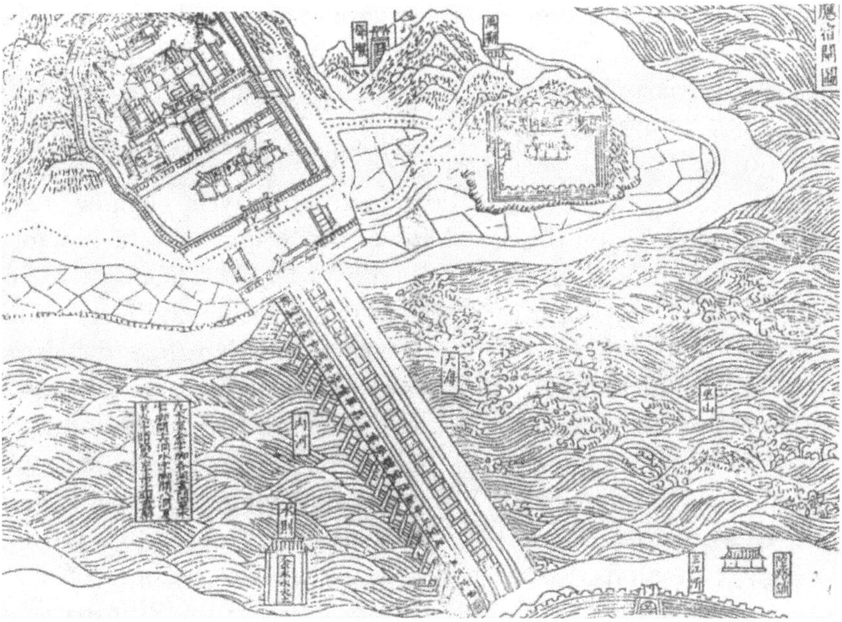

Figure 8.19: Sanjiang Sluice and Stage-Gauging Stelas, Drawn in the 19th Century

Because of riverbed aggradation on the south bank of the Qiantang Bay between the end of the Ming Dynasty and the beginning of the Qing Dynasty, Sanjiang Sluice became increasingly far from the coast. Gradually, the sluice found it was unable to play a significant role. In 1977, a new sluice, named 'New Sanjiang Sluice', was built 2.5 km north of the old Sanjiang Sluice to replace the functions of the old Sanjiang Sluice.

8.3.4 *Characteristics and Analysis of Projects*

8.3.4.1 *Distribution, Types, and Characteristics of Control Projects*

Historically, the control projects along the Zhedong Canal were located mostly at the confluences of the canal and the natural rivers, and weirs predominated in terms of the type of the projects. In terms of spatial distribution, control projects going from west to east on the Zhedong Canal are: Xixing Weir (in the Qiantang River); Qianqing Weir (in the Xixiao River); Dusi Weir (in the Jianhu Lake); Cao'e Weir (in the Cao'e River); Lianghu Weir (in the Cao'e River); Tongming Weir (in the Yuyao River); Large West Weir (in the Yuyao River); Small West Weir (in the Yuyao River); and Yiting Weir, Henghe Weir, and Doumen Weir on the north line of the Shangyu reach. In addition, the control projects on the south line of the Shangyu reach (which existed only for a short period in late Ming and early Qing dynasties) were all weirs. From the Southern and Northern dynasties to the end of the Qing Dynasty, the control projects in the Zhedong Canal with weirs as their main feature were used for as many as 1,500 years. During the operational periods of weir projects such as Qianqing Weir, Cao'e Weir, Lianghu Weir, Tongming Weir, Large West Weir, and Small West Weir, no sluices were rebuilt. It was only during a few periods that projects such as small sluices appeared, and they were only used for a short time. It was only Xixing Weir that was converted into a sluice for a significant period of time, from the Northern Song Dynasty to the end of the Qing Dynasty. However, in the late Ming Dynasty, the main function of the sluice was drainage instead of navigation.

8.3.4.2 *Analysis*

A distinctive characteristic of the Zhedong Canal is that the control projects with weirs as their main feature were operated for a longer time than any other period in history. This results from the topography and water resources of the Zhedong Canal.

The Zhedong Canal has adequate water sources and its longitudinal profile is gentle. The major issue of the Zhedong Canal was the control of water level and quantity at the confluences of the canal and natural rivers. As a result, all the control projects were located at places where the canal joined natural rivers. Therefore, in terms of the spatial distribution of control projects, the Zhedong Canal was different to the Huitong Canal, along which there was a control project approximately every 5 km.

In terms of functions, the control projects along the Zhedong Canal were operated to keep the canal from being affected by the tide and to maintain relatively stable conditions for navigation. Weirs can completely separate the canal from natural rivers and thus reduce the impact of violent fluctuations in the water levels of natural rivers on the canal by as much as possible, something that could not be achieved by sluices with the level of technical level advancement at that time. Firstly, compared to weirs, sluices were not strong enough, and the cost of construction and maintenance of the sluices was high. Secondly, because of the frequent passage of ships through the canal, sluice gates had to be opened and closed frequently, which made it unavoidable for the water level and quantity of water in the canal to be affected by the change in the water levels of natural rivers. After the Xixing Project was converted from a weir into a sluice, the operation of the sluice resulted in serious sedimentation in the canal. Thirdly, the operation of a sluice required skilful management, otherwise the sluice could not play a significant role. This has been verified by events on the Huitong Canal and the Jiangnan Canal. In contrast, the management of a weir was relatively simple. With the hydrological conditions in Zhedong region, if sluices were used as control projects at the confluences of the canal and natural rivers, it was dangerous for ships to pass through the sluices. However, the technique of towing ships over the weirs, which was laborious but relatively safe, was already mature by the Northern and Southern Song Dynasty. The principal contradiction of the Zhedong Canal as well as the technical level of that historical period has determined the long-term use of weirs as main control projects.

8.3.4.3 *The Emergence of Control Projects Combining Weirs and Sluices*

In the Zhedong region, the main functions of the sluices were to flexibly control the storage and discharge of water in rivers and lakes, while the main function of weirs was for shipping. In other words, weirs were used for shipping, while sluices were used for storing and discharging water. The longer was called 'levee', or *Tang* in Chinese, while the shorter was called 'weir'. Both levees and weirs were used to resist the tide.

Some weirs were more than 330 m long. Such long weirs were called 'weirs' in the old local chronicles, but they are called dams nowadays. Both weirs and dams were considered to be the same type of project. By means of winches, boats were pulled over the projects — such as *Tang* and weirs — located at the confluences of tidal rivers and inland channels. Another type of project was called *Qi* in Chinese, and it was used to store fresh water when it was closed and to discharge floods when it was opened. Sluices functioned in the same way. Canals required stable water levels and amounts. In addition, it was required that the canals were connected to the rivers and lakes in major river systems. As a result, a type of project that combined sluices and weirs appeared in the Ming and Qing dynasties — one part of the project was a weir for shipping, and the other was a sluice for regulating water. This type of project still exists nowadays, examples being the Jiangkou Weir and Daxi Weir. The current projects of this type have a structure of reinforced concrete, which is the only difference between the current and ancient projects. As early as the Song Dynasty, there were small sluices near weirs, but due to the limitation of topography or other factors, weirs and sluices were far away from each other. As a result, it was necessary to excavate channels between the weirs and the sluices, so the layout of the projects was complicated and the cost of construction, maintenance, and management was high.

Later, control projects combining a weir and a sluice appeared (see Figures 8.20 and 8.21). A weir and a sluice were built next to each other across a river. The weir was operated for shipping, while the sluice was used for regulating water. The weir and the sluice did not interfere with each other. The layout of the project was compact, which reduced the cost of construction.

According to the literature, the earliest project of this type (before the Wanli Era in the Ming Dynasty) was named 'Long Weir'. The location of the Long Weir is near Changba Village of Yiting Town, Yuyao City. The north line of the Shangyu Canal started at Baiguan

Figure 8.20: Today's Tongming Weir and Sluice (2010)

Note: The weir is on the left and the sluice is on the right.

Town, flowed through Changba Village, and continued to flow east to Caoshu Bridge. The Long Weir consisted of a weir and a sluice, which were combined to cross the canal. The sluice was not closed for a long time, which resulted in a decrease in the water quantity of three lakes.

The weir was made of earth. When the sluice was closed to store water, boats had to be towed over the weir, and thus the earth weir was prone to damage. Considering the requirements of navigation and water storage, the earth dam was converted into a masonry weir and there was a strict management on the opening and closing of the sluice. This type of project was often used in the Qing Dynasty. It was once suggested that this type be used for the Zhangting Project, but that plan was not implemented. Liqi Weir in the Wenxi River, 17.5 km east of Cixi County, also used this type of project. Jiangkou Weir, or Lower Tongming Weir (see Figure 8.20), which was converted in the late Qing Dynasty, was a project of the same type according to the record of the Magistrate of Yuyao County, He Zhaokai, in the 6th year of the Guangxu Era. It reads:

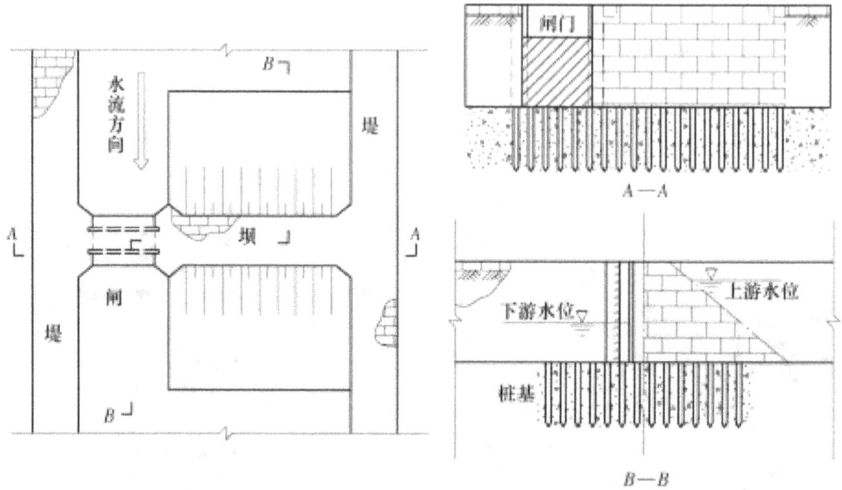

Figure 8.21: Structure of the Control Project Combining a Sluice and a Weir

> A big sluice with a sluice named Longbin next to it was built near the lower weir of Yuyao County. The weir was for shipping and the sluice for regulating water. The project played an important role in protecting the canal and resisting the tide.

That is to say, the original weir was converted into a project combining a weir and a sluice. Longbin Sluice, as mentioned above, was a small sluice operated during the Wanli Era in the Ming Dynasty. There is no way to verify when the Large West Weir was converted into a project of this type.

Chapter 9

Conclusion: Core Values of the Chinese Grand Canal

9.1 Introduction

The construction and use of the canal in China has lasted for more than 2,000 years. The values of the Chinese Grand Canal should be recognised from the political, scientific, technological, and cultural perspectives. According to the World Heritage criteria for the universal value of the canal and on the basis of research, we can summarise its core values. This summary may not fully articulate the values of the canal; nevertheless, it should provide a perspective to assess the outstanding universal value of the Chinese Grand Canal.

The canal crosses the five major rivers in eastern China, and it is an engineering system formed by a number of river reaches with distinctive local characteristics. Using the river reaches as units, this chapter provides a systematic summary of the technical characteristics of the Grand Canal (see Table 9.1).

9.2 Engineering Perspectives and Cultural Impact

Criterion I: Representing a Masterpiece of the Human Creative Spirit

The Chinese Grand Canal is a continuous waterway consisting of several artificial rivers between two natural rivers. Whether during the Sui, Tang, and Song dynasties or in the Yuan, Ming, and Qing dynasties, the Grand Canal was the backbone with the capital as the centre and access across the Haihe River, the Yellow River, the Yangtze River, the Huaihe River, Lake Taihu, and the Qiantang River. The spatial and temporal differences in water resources and the impact of flooding (especially the sedimentation and repeated waterway changes in the Yellow River) resulted in the challenges to the complex engineering and technology of the Grand Canal. Along the 2,000 km of continuous length of the Grand Canal, in each river reach, there was an independent engineering system for water regulation, water control, and other functions, including water supply and water-control projects. These engineering systems resolved the problems of interference by the natural rivers and water supply for the water-courses, and they provided navigable conditions along the canal. In terms of project planning and architecture, the Grand Canal reflected the great creativity of the Chinese people and represented the highest achievements in these areas before the Industrial Revolution in the 17th century. Some outstanding examples of the projects are listed as follows.

1. Kunming Lake and the Changhe Water-Resource Project

For the Beijing segment of the canal (the Tonghui Canal), there was no water resource from a natural river. In the Yuan Dynasty, the scientist Guo Shoujing led projects to introduce spring water into the Wengshanpo (modern-day Kunming Lake), and to construct the Changhe and water tanks of Jishuitan, Beihai, Zhonghai, and Nanhai. These projects formed reservoirs and water-supply waterways with the functions of water storage and regulation, and they provided a stable water resource for the Beijing segment of the canal. During the Ming and Qing dynasties, Kunming Lake and other artificial lakes became the Royal Garden. The water of the rivers and lakes in Beijing was supplied by the canal projects. The water system greatly impacted the urban development of Beijing from the Yuan Dynasty to the Qing Dynasty — 600 years.

2. Daicun Dam and the Nanwang Water-Diversion Project

Daicun Dam and the Nanwang water-diversion project were the key projects to cross the watershed in the Shandong section (the Huitong River) of the Grand Canal. Built in the early 15th century, the projects consisted of a dam (Daicun Dam), an aqueduct (the Xiaowen River), reservoirs (the Nanwang, Shushan, Mata, and Anshan lakes), and gates. The projects had the functions of inter-basin water transfer and seasonal water scheduling, in addition to the control of the water levels in the waterways, which provided technical support to overcome the waterhead difference of 50 m.

3. Gaojiayan Dam

Gaojiayan Dam and the resulting Hongze Lake were the key works that enabled the Grand Canal to cross the Yellow River and the Huaihe River. Gaojiayan was a masonry dam with a length of 65 km and a maximum height of 19 m. It was the longest dam among similar examples and its height held the record until the early 19th century. Gaojiayan Dam represented the highest level of Ancient Chinese dam technology. The dam was built to control the flood of the Huaihe River, to reduce sedimentation in the Yellow River, and to improve navigation conditions at the intersection of the canal and the Yellow River. These were famous river-planning adages in Ancient China — 'treating sedimentation through controlling water' and 'storing clean water to wash silt'. Both of these adages were followed in this canal project from the 16th to the 17th century. A hydrological post was set up at Gaojiayan, recording water levels continuously for 200 years. Hongze Lake was formed as a result of Gaojiayan, with a storage capacity of $35 \times 108\ m^3$ — China's fifth largest freshwater lake and the largest reservoir in the Eastern China Plain. With these projects, the Huaihe River's natural environment had undergone tremendous changes.

4. The Wu Jiangnan Project

The Wu Jiangnan levee was the key water-control project in the southern section of the Grand Canal (the Jiangnan Canal), including levees, sluices, and transport bridges. Shuize *stele* was used to control the opening and closing of the water-intake gate to achieve the regulation of the amount of water in the canal. The Wu Jiangnan project achieved water regulation in the confluence area of the canal, Lake Taihu, and the Wusong River, which allowed the Jiangnan Canal to be an independent engineering system. Shuize *stele* was the world's earliest hydrological facility on a canal.

5. Sanjiang Gate

Sanjiang Gate was the control project of the Zhedong Canal from Xiaoshan to Shaoxing and was located in the intersection of the canal, the Cao'e River, and the Qiantang River. Sanjiang Gate was built in the mid-16th century with 28 holes, a length of 103 m, and a masonry structure. The function of regulating and storing water in the canal was achieved through the joint application of two Shuize *steles* below the gates and on the canal, and through the connection with the Xixiao River. Sanjiang Gate was an important controlling project on the canal along the eastern Zhejiang coast, and it was also the largest and oldest masonry gate with multiple holes in Ancient China.

Criterion II: Reflecting the Important Exchange of Human Values in a Time Period or in a Cultural Area of the World

The water transport through the Chinese Grand Canal resulted in material exchanges between different regions, providing the channel for cultural exchanges between those regions. The Grand Canal is a continuous waterway that stretches up to 2,000 km and has become an important transport route across mainland China. During its 2,500-year life, the canal has provided the channel for economic and cultural exchanges between regions.

In the Sui, Tang, and Song dynasties, the Grand Canal crossed the Qiantang River and extended to Mingzhou (modern-day Ningbo) and to the sea. With the powerful national strength of the Tang Dynasty and the well-developed culture of the Song Dynasty, Chinese culture greatly influenced East Asia, South Asia, and other regions. The Grand Canal became an important channel for material exchange and the spread of religion between the Chinese and foreigners.

Multinational fleets came across the sea and arrived in Hangzhou, Yangzhou, and other ports, and then arrived in Luoyang, Chang'an, and other cities along the Grand Canal. Buddhist monks from Japan, Korea, and other countries, as well as Chinese monks going abroad themselves, relied on the Grand Canal's waterway. In the Song Dynasty, besides monks, a large number of foreign ambassadors, scholars, and businessmen came to China through the canal, some of whom took on year-round positions in cities along the canal. In the Yuan Dynasty, with opening up of the Beijing–Hangzhou Grand Canal, Europeans began to appear along the canal, one of whom was the famous Venetian Marco Polo. He had come to China several times, working as an official for a long period, and he travelled to many cities and towns along the Grand Canal. During the Ming Dynasty, most of the tributary envoys from South Asia landed in the coastal areas of Fujian and Zhejiang before travelling north along the canal. Besides tributes, they also carried large quantities of goods, books, and cultural relics, and they exchanged these goods along the way.

In the late Ming Dynasty, Western missionaries, with the Italian missionary Matteo Ricci as their representative, travelled along the canal and introduced Western science and technology to the Chinese people. They also introduced China's customs to Westerners back home. During the Qing Dynasty, the canal was the earliest route for transmission of foreign religions. Muslim mosques, Christian churches, and monasteries first appeared along the Grand Canal and then extended to the mainland rural and urban areas.

Criterion III: Providing Unique or Exceptional Testimony for Existing or Disappeared Civilisation or Cultural Traditions

During the 2,500 years from the Spring and Autumn and Warring States periods to the Qing Dynasty, the canal was constructed with the capital as its key destination, and the canal's management mainly focussed on the transportation of grain to the capital. These activities reflected the will of the state under different historical periods; the state had a monopoly over the rivers and the distribution of water resources. Due to the regional natural conditions, China's grain production areas were mainly located in the southeast, while the capitals of the unified empires often located in the north. The separation between the political and economic centres determined the role of the Grand Canal as a transportation lifeline, supporting the national economy. The construction and utilisation of the Grand Canal took place during the most important periods in Chinese history, testifying to the political, social, and cultural patterns of those periods.

The Chinese Grand Canal was initially built for military purposes as a series of interval canals. During the Warring States Period, the Honggou Canal was excavated to connect the Yellow River and the Huaihe River, and the Hangou Canal was excavated to connect the Huaihe River and the Yangtze River. During the Three Kingdoms Period, canals were constructed at Baigou, Pingluqu, and Licaoqu. These canals were primarily used to transport troops and provisions during wartime. Since the demise of the unified empires of Qin and Han, China entered chaos for more than 400 years. The Sui Dynasty saw the country became a unified empire again. In the Sui Daye Era (AD 605–618), the canals of Tongjiqu and Yongjiqu were constructed and Sanyangdu (the Hangou Canal in the previous dynasty) and the Jiangnan River were systematically renovated. Although the Sui Dynasty lasted for less than 40 years (AD 581–618), over the next 600 years the Grand Canal became the economic artery of the Tang and Song dynasties. The main function of water transport was to reach the capitals of Luoyang (in the Tang Dynasty) and Bianjing (in the Song Dynasty). The management system of the Grand Canal became an important component of national administration.

In the Yuan, Ming, and Qing dynasties, Beijing was the capital. Between the political centre in the north and the economic centre in the south, a more direct route was selected for the new Grand Canal. This new route crossed the Yellow River at the junction of the Yellow River and the Huaihe River successfully, crossing the Shandong horst with a terrain elevation difference of over 50 m, which resolved the water-source problem of the Beijing section with a steep slope. This canal route was the genesis of the Beijing–Hangzhou Grand Canal and is still in use today.

The canal in the Sui, Tang, and Song dynasties and the canal in the Yuan, Ming, and Qing dynasties were operating during the most important periods of China's dynasties. In order to achieve transportation along the canal, the states invested hundreds of thousands of man-hours and a tremendous amount of material resources in the construction and management of the canal. Through national legislation and institutions and mandatory water regulation, the canal's water resources were protected. It became a link between the grain-producing areas and the consuming areas, connecting the political and economic centres. The canal operations promoted the exchange of goods in different regions, thereby enabling the formation of a number of cities and towns along the canal, affecting the lifestyles and customs of the residents, and contributing to the rich local culture. The construction and operation of the Grand Canal reflected China's political, social, and cultural patterns across these historical periods, providing a vivid testimony for 2,500 years of China's history.

9.3 Summary of Heritage Based on the Canal Reaches and Key Engineering Remains

After being affected by the Yellow River moving south and crossing the Huaihe River for 700 years, the Tongjiqu Canal (or Bianhe Canal) and the Yongjiqu Canal, which had had an important influence on Chinese history, were basically buried with sediment and wasted away. The canals still currently existing — the Beijing–Hangzhou Canal and the Zhedong Canal — continue to offer significant benefits. Since the cultural values of the Grand Canal have been ignored for a long time and the heritage of canal engineering has been continuously damaged, it is urgent to protect some sections of the canal and the ruins along it. Through a survey and investigation of the ruins of the projects, we have summarised the key engineering projects and their technical characteristics (see Table 9.1). We hope the canal's sections and key engineering projects will be protected.

Table 9.1: The Key Engineering Projects on the Grand Canal and their Technical Characteristics

Canal section	Project type	Key works	Value analysis
Tonghui Canal	Water-source project	The Tonghui Canal above Shichahai was a water-source project that included water-introducing engineering from the Beijing Yuquan Mountain, Changhe water delivery engineering, Kunming Lake, and Shichahai. Artificial lakes were formed through the water-source project to achieve annual water regulation of the Tonghui Canal and effectively guarantee water supply to the canal.	This water-source project was an outstanding example of canal planning. The project provided a guaranteed supply of water for the Tonghui Canal and also performed municipal and environmental functions. The rivers and lakes established by the project helped to create the beautiful landscape of ancient Beijing.
	Sluice group	In the Tonghui Canal, 24 regulating sluices were used to control water and ensure adequate depth for navigation. The only remaining sluices include Guangyuan, Gaoliang, Chengqing, and Qingfeng.	The sequential placement of the sluice group enabled boats to sail against the current from Beijing to Tongzhou, where the terrain elevation difference was 20 m. They also helped to save water. The Tonghui sluice group was a typical engineering project on the ancient canal and represented an outstanding project-planning achievement.
South and North Canals	Artificial bend	Through the design of the curvature of the canal, the canal's longitudinal gradient was reduced by 50 per cent so that the canal's terrain-elevation difference was decreased and the waterway's navigation conditions were	With minimal engineering effort, the curvature design provided better navigation conditions and security for boats travelling along the canal. The principle of 'three bends to replace one gate' was another

		improved. The curvature design slowed the pace of floodwater in the canal's curved segments to enhance the safety of navigation.	characteristic of the planning of the Grand Canal's waterways.
	Water-drainage sluice	To solve the drainage problem in the flood season and the water-supply problem in the dry season for the South and North Canals, water-drainage sluices were constructed at the east bank of the canal and Jianhe was excavated to drain floodwater to the east and into the sea. The main sluices (dams) to control water drainage included Hexiwu, Qinglongwan, Kuangergang, Jiedi, Xingji, Machang, Shaomaying, and Sinvsi. The only remaining sluices include Kuangergang, Jiedi, Machang, Shaomaying, and Sinvsi.	The rational arrangement of Jianhe and the sluices met the comprehensive needs for the canal's water regulation and flood control. The combination of the dams and gates reflected an outstanding achievement in the planning, design, and construction of the ancient hydraulic engineering projects.
Huitong River	Water-source and diversion projects	The water-source and diversion projects and the connection of the projects with the natural river in the northern Jining section of the Huitong River were conceived alongside the key water projects of Linqing, Nanwang, and Jining. Each key water project consisted of the water doumen (or dam), water tanks, sluices, and other components. The main remains include the four gates (Huitong, Linqing, Zhuanzha, and Banzha) of the Linqing key	These projects effectively solved the water scarcity problem on the Huitong River. Boats could smoothly cross the Shandong horst with a terrain elevation difference of 50 m. The projects also allowed the Beijing–Hangzhou waterway to enter full operation. The Daicun Dam — Nanwang's key water project — represented the highest achievement in terms of engineering technology on the Chinese Grand Canal.

(*Continued*)

Table 9.1: (*Continued*)

Canal section	Project type	Key works	Value analysis
		project; Daicun dam, the Xiaowen River, and the Nanwang gate of the Nanwang key project; and Gangcheng dam, Jinkou dam, and the Tianjing gate of the Jining key project.	
	Lakes and water tanks	The several lakes in the south of Jining (Nanyang, Zhaoyang, Dushan, and Weishan) were the result of the Yellow River dike breaking and the resulting engineering measures. These lakes (or water tanks) had the functions of flood storage and detention, supplying the canals with water, and improving the quality of navigation along the waterways.	These lakes formed a unique canal waterway and had good ecological and landscape values.
	Sluice group	The sluice gates of the Huitong River were important projects to ensure the Grand Canal could cross the Shandong horst. There were 31 gates in the Yuan Dynasty and 40 gates in the Ming Dynasty, and thus the Huitong River was known as the 'gate river'. These gates still exist today.	The combined scheduling of the gates effectively solved the terrain-elevation differences and lack of control over the water, thereby improving the navigation conditions.

The Middle Canal (Zhong Canal)	Transportation hub (joint)	The position of the joint between the Yellow River and the canal in the north of the Yellow River constantly changed with the river's sedimentation. After the Middle Canal was built, the joint moved to Yangzhuang. There are remains of the joints along the left bank of the Yellow River.	The remains of the joint positions and related engineering works are proof that the Chinese struggled with the Yellow River, which had the highest concentration of sediment out of any river in the world.
	Water tank	The artificial lakes were the result of the influence of the Yellow River's sedimentation and the engineering measures and performed the functions of water supply and water regulation. Currently, the Luoma and Huangdun lakes are still used.	This water tank testifies to the hard work undertaken to control the influence of the Yellow River.
Huaiyang Canal	Transportation hub (joint)	Qingkou joint was located in the intersection of the Beijing–Hangzhou Canal, the Yellow River, and the Huaihe River, which was the most complex region along the canal. In the Ming and Qing dynasties, the strategy of 'storing clean water to wash silt' was implemented. The key project of the Qingkou–Hongze Lake was built and used to provide support for boats to cross the Yellow River and the Huaihe River. Current engineering remains include Gaojiayan (the Hongze lake levee), the iron buffalo for water control, Matouzheng wharf, Yuhuang dam, and hydraulic piles.	The transportation hub had been used for 300 years to deal with the frequent flooding of the Yellow River. In the Qingkou region, all the traditional methods of hydraulic construction were used; among those, the layout and structure of the Gaojiayan dam represented the highest achievement in water-conservancy technology in China before the 17th century.

(*Continued*)

Table 9.1: (*Continued*)

Canal section	Project type	Key works	Value analysis
	Water tank	The water tanks were the artificial lakes formed by engineering measures and included the Hongze, Gaoyou, and Baoying lakes. The water tanks were used for the regulation and storage of water and the control of water levels in the canal.	The water tanks effectively regulated the amount of water in the canal and ensured sufficient depth for smooth navigation.
	North transportation hub	The Guazhou port located on the north shore of the Yangtze River and was a major transportation hub for the waterway from the Yangtze River to the canal. After the collapse of the north shore of the Yangtze River, the ancient Guazhou port no longer existed, but there is still a section of the canal that connects with the river.	The north transportation hub was an important landmark reach of the canal.
Jiangnan Canal	South transportation hub	Historically, the joints between the Jinagnan Canal and the Yangtze River included Jing Kou, Mengduhe Kou, and Jiangyinxia Gang. These key water projects performed the functions of tide introduction, water conservation, and navigation. The Jing Kou gate lock, built in the Northern Song Dynasty, was the most famous of these. Existing remains include the Zhenjiang Jing Kou gates and canal warehouses.	The south transportation hub was an important landmark reach of the canal.

Water tank	Lianhu Lake was the water-source project in the Zhenjiang section of the north portion of the Jiangnan Canal, being operated according to the adage that 'releasing an inch of the lake water made the canal water rise a foot'. This allowed the normal operation of the canal segment from Changzhou to Zhenjiang to continue. Existing remains include the Lianhu Lake sluice.	Lianhu Lake is the sole proof of the water tanks on the Jiangnan Canal.
Sluice group	In the Tang and Song dynasties, the world's earliest locks — complex gates — were built between Zhenjiang and Wuxi and between Jiaxing and Haining. Among the complex gates, the Jing Kou gate, Lucheng gate, Benniu gate, and Chang'an gate were the most famous. Through joint use of the upper and lower gates, boats could overcome different water level elevations to go through the gates. Chang'an gate is the only remaining complex gate.	The complex gate positions were the landmarks of the canal during the Tang and Song dynasties, demonstrating the technical achievements on the Grand Canal during these periods.
Key project combined with the natural water system	The key project was represented mainly by the Wujiang Lu project of Suzhou, which included towpaths along the levee, drainage works (culverts), Shuize *steles*, and bridges. The completion of the Wujiang Lu project in the Song Dynasty separated the Jiangnan Canal	The Wujiang Lu project was an indication of the completeness of water-conservancy engineering on the canal during the North Song Dynasty. The Shuize *stele* was the earliest water-level-measurement facility in China.

(*Continued*)

Table 9.1: (*Continued*)

Canal section	Project type	Key works	Value analysis
		from the Tai Lake and the Wusong River. The bridges and culverts were used to provide the canal with water from Tai Lake and drained floodwater to the Wusongj River, thus ensuring stable water levels along the canal.	
	Moat	The moats in Wuxi, Suzhou, Jiaxing, and Hangzhou were once the main canal route performing the functions of traffic organisation and water regulation and storage. After the further development of the canal, they became branch lines and were kept as a complete engineering system.	The moat sections comprised a complete engineering facility and produced a unique water landscape. The moat sections and the street bear witness to the exchange of materials and cultures along the canal.
Eastern Zhejiang Canal	Water-source project	Represented by Sanjiang gates, this water-source project with 28 consecutive gates achieved water control of the canal section between Xiaoshan and Shaoxing. When the canal's water levels were low, the gates were opened to introduce water from the Cao'e River to the canal; when the canal's water levels were high, during the flood seasons, the gates were opened to discharge canal water. Two Shuize *steles* were set up in Shaoxing City and at Sanjiang gate. The gates were opened or closed according to the water levels within the canal.	Sanjiang gates formed the earliest multiple-hole sluice in Chinese history. The canal's water levels were controlled to achieve quantitative control over the canal's water.

Regulation project	This type of project was a combination of dams and gates as best represented by the Tongming and Daxi dams. The regulation project was the engineering solution of joining the natural rivers and canalised waterways, performing the functions of tide blocking, supplying water to the canal, and maintaining an adequate water level to allow navigation along it.	The Tongming and Daxi dams, built in the Song Dynasty, were typical water projects on the Eastern Zhejiang Canal. These projects reflected that the planning concepts and architectural styles could harmoniously coexist with the rivers.
Moat	Moats performed the functions of traffic organisation and water regulation and storage. The moats in Shaoxing, Yuyao, and Ningbo were especially useful in this respect. Through Shuize *steles*, the water levels of the Shaoxing and Ningbo moats were controlled quantitatively.	The moat sections comprised a complete engineering facility and produced a unique water landscape. The moat sections and the street bear witness to the exchange of materials and cultures along the canal.

Index

www.ingramcontent.com/pod-product-compliance
Lightning Source LLC
LaVergne TN
LVHW010055220926
848484LV00005BA/1082
* 9 7 8 1 9 4 5 5 5 2 0 3 8 *